Rimeh ISMAIL

Silver catalysts for toluene oxidation

Rimeh ISMAIL

Silver catalysts for toluene oxidation

Synthesis and characterization of silver-based catalysts for toluene oxidation

ScienciaScripts

Imprint

Cover image: www.ingimage.com

This book is a translation from the original published under ISBN 978-620-6-71554-2.

Publisher:
Sciencia Scripts
is a trademark of
Dodo Books Indian Ocean Ltd. and OmniScriptum S.R.L publishing group

120 High Road, East Finchley, London, N2 9ED, United Kingdom
Str. Armeneasca 28/1, office 1, Chisinau MD-2012, Republic of Moldova, Europe
Managing Directors: Ieva Konstantinova, Victoria Ursu
info@omniscriptum.com

Printed at: see last page
ISBN: 978-620-8-56371-4

Contents

TO MY DEAR PARENTS

No dedication can express my respect, my eternal love and my consideration
consideration for the sacrifices you have made for my education
my education and well-being.
Thank you for all the support and love you've given me since I was a child.
and I hope that your blessing will always be with me.
May this modest work be the fulfilment of your much-vaunted wishes, the fruit
of your countless sacrifices, even though I will never pay you for it
enough.
May God, the Most Powerful, grant you health, happiness and long life and
never let you down.

To my dear, adorable brothers and sisters

Who have always supported and encouraged me throughout years of study.
I wish you a life full of happiness and success, and may Almighty God protect and guard you.
protect you and keep you.

To my dear uncle Tarek

My advisor, and loyal friend, who helped me through difficult times.
I can never thank you enough for your kindness, your generosity and your invaluable help.
help.

To my dear cousins " Adel, Monia , Samira , hanen , noura and mouna "

Who have helped me through difficult times and taken me gently by the hand
to get through difficult times together....
I can't thank them enough for their endless help and encouragement.
and encouragement. Thank you so much.

TO MY DEAR LITTLE NEPHEWS AND NIECE

"Yaakoub, Abd Rahmen and khadija

No dedication can express all the love I have for you.
your joy and cheerfulness fill me with happiness.
May God watch over you, enlighten your path and help you to achieve your most
your most cherished dreams.

To my dear brothers-in-law Yessin and Nassim

Who have always stood by me through thick and thin.
I can't thank them enough for their support, their sound advice and their invaluable help.
and their invaluable help, thank you very much.

To my friends

To all my loved ones

Please find in this work the expression of my deepest respect
and my most sincere affection.

RimehISMAIL

Acknowledgements

*First of all, I would like to thank **God** for giving me the health, strength, willpower and patience to be able to continue my studies and complete this work.*

*This work was carried out in the Chemistry Department of the Tunis Faculty of Science, in the Chemistry of Materials and Catalysis Laboratory directed by Mr **Zouhaier KSIBI**, Professor at the Tunis Faculty of Science. I would like to express my sincere thanks and gratitude to him for welcoming me in his team and for directing this thesis.*

*I would like to express my deep gratitude to Mrs **Jihène ARFAOUI**, Assistant Professor at the Faculty of Science in Tunis, for having supervised me during this thesis and for having closely followed all of my work. I would also like to express my deepest respect for her patience, kindness, availability, her pertinent remarks and advice, her very instructive meetings and her invaluable help.*

*I would like to express my warmest thanks and sincere gratitude to Mr **Abdelhamid GHORBEL**, Honorary Director of the Chemistry of Materials and Catalysis Laboratory and Professor at the Tunis Faculty of Science, for his human qualities, his infinite help, his encouragement and his invaluable advice. I would like to express my deepest respect to him.*

*My warmest thanks go to Mr **Gérard DELAHAY**, CNRS Research Director at the Ecole Nationale Supérieure de Chimie de Montpellier (ENSCM), for his warm welcome, his kindness, his availability and his help, particularly with the catalytic set-up during my internships at the Laboratoire de Matériaux Catalytiques et Catalyse en Chimie Organique in Montpellier.*

*My most sincere thanks also go to Mr **Noureddine AMDOUNI**, Dean and Professor at the Tunis Faculty of Science, for the honour he has done me by agreeing to chair the jury for this thesis.*

*I would like to warmly thank Mr **Nizar BELAKHAL**, professor at the Institut National des Sciences Appliquées et de Technologie (INSAT), and Mrs **Noura FAKHAR BOURGUIBA**, lecturer at the Faculté des Sciences de Tunis, for agreeing to examine and judge this work, as rapporteurs.*

*I would also like to thank Mr **Mohamed Faouzi ZID**, professor at the Tunis Faculty of Science, for agreeing to take part in the jury for this thesis as an examiner.*

Not forgetting to thank all the "old" and "new" members of the Materials Chemistry and Catalysis Laboratory at the Tunis Faculty of Science.

Thank you all for your help and support over the years....

General introduction

The release of atmospheric pollutants, including volatile organic compounds (VOCs) such as aromatics (toluene, xylene, etc.), ketones (acetone, methyl ethyl ketone, etc.), alcohols (methanol, ethanol, propanol, etc.), acetates (ethyl acetate, etc.) and chlorine derivatives (methylene chloride, perchloroethylene, etc.), is a growing and regulated concern. The toxic properties of VOCs and their harmful effects on ecosystems and humans [1] have led to the introduction of increasingly stringent regulations to control and reduce the concentrations of these pollutants in gaseous emissions.

The main ways of treating atmospheric pollutants are based on physical matter transfer such as absorption, adsorption and condensation [2]. Adsorption of VOCs, for example, enables these pollutants to be transferred from the air to a porous solid medium such as activated carbon without destroying them, which generates toxic solid waste (when the VOCs captured cannot be recycled) [3,4] and therefore only corresponds to the transfer of pollutants from one phase to another.) So-called "destructive" techniques are used to solve this problem. Among these, thermal combustion (or incineration) is the most widely used, but its energy cost is high because it only works at high temperatures (> 1000°C) and leads to the formation of NOx and by-products that are more toxic than the original compounds (dioxins, furans, etc.) [3].

Catalytic oxidation of VOCs is a more attractive technique than incineration because, in the presence of a catalyst, it enables total elimination of these compounds at low temperatures (< 500°C) while avoiding the formation of NOx and toxic by-products [5]. The literature has shown that noble metals such as palladium and platinum are very active in the catalytic oxidation of VOCs. However, because these metals are expensive, they are increasingly being replaced by transition metals, which are also very active in the oxidation of many VOCs. Although silver-based catalysts have shown good catalytic activity in the elimination of various VOCs, little work has been reported in the literature on the use of this type of solid in the toluene oxidation reaction. In this work, we were interested in the preparation of new silver-based catalysts supported on ZrO2 and M-ZrO2 solids (with M = Ce or Fe) for the toluene oxidation reaction. The effect of Ag and M contents on the physicochemical properties of these solids and on their catalytic activity in toluene oxidation is studied.

In the first chapter of this manuscript, we report on a bibliographical study of volatile organic compounds (VOCs, in particular toluene), processes for treating these compounds (in particular catalytic oxidation and the various catalysts used), and the sol-gel method (used in this work for preparing catalyst supports).

The second chapter presents the experimental techniques, equipment and methods used for the preparation and characterisation of the samples (ZrO2; M- ZrO2; Ag/ZrO2; Ag/M-ZrO2). The results of the analysis of the physico-chemical and catalytic properties of the samples obtained are presented in the third and fourth chapters respectively. In the general conclusion we summarise the main results obtained in this work.

Chapter I

Bibliographical study

1. 1. Introduction

This chapter presents a review of the literature on volatile organic compounds (VOCs), which are air pollutants that are harmful to humans and the environment, and on VOC treatment processes, in particular the catalytic oxidation of toluene and the various catalysts used. This chapter also includes a presentation of the sol-gel method, used for the preparation of catalytic supports, and a justification for the choice of silver-based catalysts used in this work for the oxidation of toluene.

1.2. Volatile organic compounds (VOCs)

1.2.1. Definition and chemical properties of VOCs

Volatile organic compounds (VOCs) are considered to be the main atmospheric pollutants [6-7]. They are formed essentially from carbon (C) and hydrogen (H), but may also contain oxygen (O), nitrogen (N), halogens (fluorine (F), chlorine (Cl), bromine (Br), iodine (I)), sulphur (S), phosphorus (P) and silicon (S).

According to the European Union: A VOC is any organic compound with a vapour pressure greater than or equal to 10 Pa at 20°C [8]. It is characterised according to its physical and chemical nature, its residence time in the atmosphere, its boiling point and its saturation vapour pressure. According to the US Environmental Protection Agency (EPA), a volatile organic compound is defined as any organic product with a sufficiently high vapour pressure (at room temperature) to be almost completely in the vapour state. The EPA also specifies that it is a product whose saturation vapour pressure is greater than 13.3 Pa at a temperature of 25°C and which has sufficient longevity and reactivity with the atmosphere to be able to take part in photochemical reactions [8].

Depending on their chemical composition, VOCs can be classified into four main families:

- *Oxygenated VOCs*: alcohol (ROH), aldehydes (RCOH), ketones (RCOR'), esters (RCOOR'), etc.
- *Saturated or unsaturated hydrocarbons*: alkanes (C_nH_{2n+2}), alkenes (C_nH_{2n}), alkynes (C_nH_{2n-2}).
- *Mono- and polycyclic aromatic hydrocarbons (MAH and PAH):* benzene (C_6H_6), toluene (C_7H_8), xylene (C_8H_{10}), anthracene ($C_{14}H_{10}$), etc.
- *Halocarbons:* dichloromethane (CH_2CL_2), chloroform ($CHCL_3$), dichloroethylene ($C_2H_2Cl_2$).

1.2. 2. Main sources of VOC emissions

There are two types of VOC emission source: anthropogenic and biogenic [7, 9-12].

- *Anthropogenic sources* include emissions linked to human activities and are subdivided into three sectors: industrial, residential and transport.
- *Biogenic sources* include natural emissions from animals, vegetation and forest fires.

I.2.3 Impact of VOC emissions

Volatile organic compounds are a major source of air pollution. They have direct harmful effects on human health, linked to their toxicity and odour, and indirect effects on the environment, such as smog and global warming [9-10,13-15].

I.2.3.1. Direct effects of VOCs: on human health

VOCs, particularly oxygenated ones, are harmful to humans, leading to allergic reactions such as asthma [16,17] and even cancer [18,19]. Their toxic effects and the disorders they can cause vary from one compound to another, and are presented in Table 1.

Table 1. Toxic effects and disorders caused by VOCs.

VOCs	*Effects and problems caused*
Most VOCs	Headaches
Halogenated or aromatic hydrocarbons	Skin irritation
BTEX *, substituted benzene rings and aldehydes	Eye irritation
Diisocyanate and aromatic hydrocarbons	Irritation of the respiratory organs
Toluene, chloroform and methyl chloroform	Heart problems
Benzene, toluene and halogenated hydrocarbons	Digestive disorders
BTEX*, cumene and halogenated hydrocarbons	Kidney disorders
Aromatic and halogenated hydrocarbons	Nervous system disorders
BTEX*, styrene, alkenes, halogenated hydrocarbons and formaldehydes	Carcinogenic and mutagenic effects

With : BTEX* = Benzene, Toluene, Ethylbenzene and Xylene

I.2. 3.2. Indirect effects of VOCs: on the environment

Volatile organic compounds have a number of impacts on the environment. Under the action of solar radiation, these compounds can release oxidising species that influence our climate [20]. Their toxicity has a direct effect on flora and fauna and they are partly responsible for the formation of tropospheric ozone (O_3), which is considered to be a greenhouse gas with an effect 2000 times greater than that of CO_2 [20]. The formation of this type of ozone is initiated by solar radiation in the trosphere [21] and is accompanied by the production of other species with acidic and oxidising properties (organic nitrates, citric acid, aldehydes, hydrogen , etc.). These contribute significantly to the acidity of rainwater due to their high solubility.

A classification of VOCs into three families according to their role in tropospheric ozone production is shown in Table 2.

VOCs, particularly chlorinated ones such as trichloroethane ($C_2H_3C_{13}$) and carbon tetrachloride (CC_{14}), play a direct role in global warming by contributing to the depletion of stratospheric ozone [22]. Their concentration in the upper layers of the atmosphere leads to warming (a temperature increase of 2 to 5°C), which in turn affects the climate and rainfall.

Table 2. Classification of VOCs according to their role in the production of tropospheric ozone [23].

VOCs	**Role in tropospheric ozone production**
Aromatics except benzene Alkenes Alkanes > C6 (except dimethylpentane) Aldehydes (except benzaldehyde) Natural (isoprene)	Quite important
Alkanes, Ketones, Alcohols, Esters	Not important
Alkanes (methane, ethane) Alkynes (acetylene) Aromatics (benzene) Aldehydes (benzaldehyde) Ketones (acetone) Alcohols (methanol) Esters (methyl acetate) Chlorinated	Very small

hydrocarbons (methyl chloroform, trichloroethylene and tetrachloroethylene)	

I.2.4 VOC treatment processes

There are two categories of VOC treatment processes: recovery processes and destructive processes. Their operating principles as well as their advantages and disadvantages are detailed in the following paragraphs.

I. 2. 4. 1 VOC recovery processes

VOCs can be recovered by condensation, adsorption, absorption or membrane separation (table 3).

Table 3. Recoverable VOC treatment processes.

Technical	***Operating principle***	***Benefits***	***Disadvantages***	***Ref. Biblio.***
Condensation	-Transforming gaseous VOCs into liquid or solid form by lowering the temperature. They are then treated for reuse or destruction.	*Recovery of condensed VOCs. *Reliability and flexibility of use. *Suitable for very high concentrations (>5000ppm).	*Limited to VOCs with a boiling point < 40°C. *The presence of dust in the effluent can lead to the formation of liquid or solid vesicles that are difficult to separate mechanically.	[24-25]
Adsorption on activated carbon and zeolites	Based on the ability of solid surfaces to bind certain molecules reversibly, through weak Van der Waals-type bonds.	*95% efficiency. *Easy to use. *Possibility of recovering pollutants.	*Periodic maintenance. *Regeneration is expensive. *Very restrictive implementation conditions. *This phenomenon is exothermic	[26-29]
Absorption	Transfer of mass (solute) from the vapour phase to the liquid phase by washing with an absorbent (solubilisation).	*Selective treatment. *Efficiency of up to 98%.	*High cost. *Absorbent is sometimes toxic.	[30-33]
Membrane separation	Fractionating a gas mixture in contact with a semi-permeable membrane to which a pressure differential is applied.	*Operational up to temperatures in excess of 200°C. *Recovery rates of up to 99%.	*High running costs.	[34-35]

I.2.4.2 Destructive processes for treating VOCs

Of the destructive processes (listed in Table 4), thermal oxidation or incineration is the most widely used, but it is energy-intensive because it operates at high temperatures (>1000°C) [36]. In addition, it can lead to the formation of by-products that are more toxic than the initial compounds. However, catalytic oxidation is proving to be an interesting technique because, in the presence of a catalyst, it enables VOCs to be destroyed at low temperatures, which has the advantage of reducing the economic cost compared with incineration and avoiding the formation of by-products that are even more toxic than the initial compounds [37-39]. In addition, this reaction enables VOCs and carbonaceous particles to be converted selectively into carbon dioxide (CO_2) and water (H_2O) [11-12].

Table 4. Destructive processes for treating VOCs.

Techniques	***Principles***	***Benefits***	***Disadvantages***	***Ref. Biblio.***

Biological destruction	Biodegradation in an aerobic aerobic environment of pollutants by certain micro-organisms.	*Low running costs. *Can be operated at room temperature.	*Requires constant monitoring of the system (humidity, temperature, etc.). *Limited to biodegradable VOCs.	[40-41]
Thermal incineration	Destruction of gaseous pollutants with a flame in the presence of O2 (combustion).	*99.99% destruction efficiency. *The heat released can be recovered.	*NOx formation. * Very high energy costs. * Risk of explosion * Delicate for halogenated VOCs.	[42]
Catalytic oxidation	Degradation of hydrocarbons into CO_2 and H_2O over a catalyst at lower temperatures than incineration.	*Low destruction temperatures (200-400°C). *No NOx formation. *Highly effective: up to 99% destruction.	*Risk of catalyst deactivation or poisoning. *Precious metal catalysts are too expensive.	[34]

I.2.5 Catalytic oxidation of VOCs

Heterogeneous catalysis has been adopted as an effective means of destroying VOCs at relatively low temperatures (<500°C). It requires the presence of a solid chemical entity containing an active phase and enabling organic reactions to be carried out under often mild conditions [43]. In this case, the reaction takes place at the gas/solid or liquid/solid interface and proceeds in five stages [44], as shown in the diagram below:

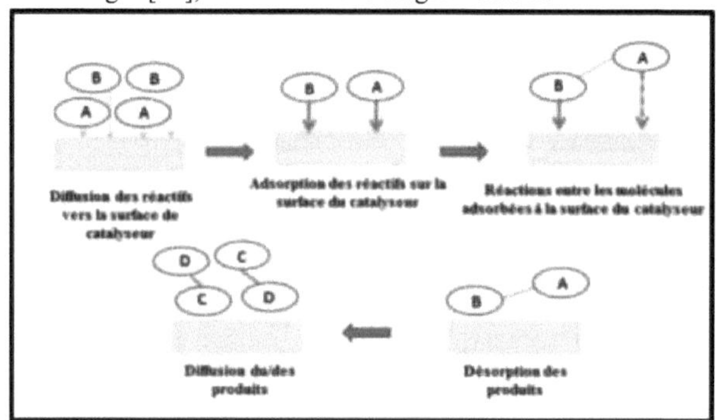

Stages of a chemical reaction in heterogeneous catalysis.

<u>Stage 1</u>: Diffusion of reagents from the fluid phase to the surface of the catalyst.

<u>Stage 2</u>: Adsorption of the reagents onto the catalyst surface: the reagent molecules attach themselves to the catalyst surface by physical adsorption (physisorption) or chemical adsorption (chemisorption).

It should be noted that physisorption is a physical phenomenon of adsorption which occurs at low temperatures between the reactant and the catalyst through weak interactions of the Van Der Waals type (dipole-dipole interactions of energy less than 20 kJ/mol). Chemisorption, on the other hand, is a chemical phenomenon that occurs at slightly higher temperatures than physisorption. In this case, a strong interaction of the covalent bond type is formed between the reactants and the catalyst, with the adsorbed molecules becoming much more reactive than in the fluid. Although adsorption of the reagents is essential, it must not be too strong, as there is a risk of forming species that are too stable, preventing their subsequent reaction/desorption and thus reuse of the catalytic site.

Stage 3: Chemical reactions on the surface of the catalyst.

Stage 4: Desorption of the reaction products, which are detached from the catalytic surface in an endothermic process. This stage is slower in the case of chemisorption than in the case of physisorption, because it is much more energetic.

Stage 5: Diffusion of the product(s) into the gaseous reaction medium.

The catalyst is finally regenerated and ready for a new catalytic cycle.

1.2. 5.1. Catalysts used for the oxidation of VOCs

An ideal catalyst for the oxidation of VOCs must be active at low temperatures, thermally stable, possess high co2 selectivity, be inexpensive, durable, capable of efficiently oxidising low concentrations of pollutants at high flow rates with little or no deactivation and must show good resistance to poisoning. The literature reports several studies involving the use of different types of bulk or supported catalysts in the oxidation of different VOCs.

Mass catalysts are metals or metal oxides in a simple or mixed state. They are characterised by the homogeneity of their surface composition and their volume (their internal part). These solids are generally characterised by low porosity and specific surface area, which limit contact with the reactant(s).

Supported catalysts are formed by an active phase (responsible for the reaction), generally a noble or transition metal, deposited on a solid support. The choice of support depends on several criteria:

- Its structural stability under reaction conditions.
- Its own catalytic activity is linked to the presence of acid-base functions or redox couples.
- Its large specific surface area, which often allows good catalytic activity due to the increased reactant-catalyst contact surface.

Some examples of the mass and supported catalysts and VOCs treated are given in Tables 5 and 6.

Table 5. Mass catalysts used in the oxidation of VOCs.

VOCs treated	***Catalysts used***	***Main results***	***Ref. biblio.***
Propane and toluene	CO3O4	-Active at low temperatures (160°C for propane and 200°C for toluene). -CO_2 selective: 100% for propane and 99% for toluene.	[45]
Chlorinated VOCs : Dichloromethane, -1,2-dichloroethane - Carbon tetrachloride - *Trichloroethylene* - *Tetrachloroethylene.*	CeO2	- CeO_2 is characterised by a high oxygen transport capacity and an ability to transform easily between its reduced and oxidised forms (Ce^{3+}-Ce^{4+}). - 90% destruction at 260°C for tetrachloroethylene and 160°C for dichloromethane. - The presence of water (30,000 ppm) inhibits the destruction of trichloroethylene.	[46]

Propene and propane	МП2О3, Fe_2O_3,Mn3O4 Мп2О3-Ре2О3	Catalytic activity decreases in the following order: Mn2O3-Fe2O3>Mn3O4>Mn2O3> Fe2O3	[47]
1,2-dichloi'()ethane	CO3O4	-Co_3O_4 shows strong potential catalytic converter. - 100% $CO_{(2)}$ selectivity is obtained.	[48]
Propane and propene	MgCr2O4	-Active in the 173-473°C temperature range. -CO_2 selective.	[49]
Benzene	CuO-CeO2	-Catalytic activity is linked to the presence of reactive surface species $Cu^{(2+)}$) and O_2, the formation of which is favoured by the increase in calcination temperature.	[50]
Light hydrocarbons	Co3O4,СиО ,NiO, МП2О3, Fe2O3 СГ2О3.	- Co_3O_4 is the most active. -Activity decreases in the following order: Co3O4>CuO>NiO> Mn2O3>Fe2O3> Сг2О3.	[51]

Table 6. Supported catalysts used in the oxidation of VOCs

VOCs treated	Catalysts used	Main results	Ref. Biblio.
Carbon monoxide	Pd/AlO Pd/CeO2/Al2O3	- The nature of the Pd precursor influences the catalytic activity: catalysts prepared using $PdCl_2$ and $Pd(acac)_2$ show better activities than those prepared using $Pd(NO_3)_2$. -The presence of CeO_2 promotes the oxidation of CO.	[52]
Benzene	Pt/y-AlA Pt/h-BN	- The nature of the support influences catalytic activity: Pt/h-BN is more active than Pt/γ-Al_2O_3.	[53]
O-xylene	Pt/zeolite	-The Pt content influences catalytic activity: 0.3% Pt/zeolite is active from 225°C, while 0.1% Pt/zeolite is active at around 240°C.	[54]
Carbon monoxide, ethanol and toluene	Pt/TiO_2, Pd/TiO2 Rh/TiO2, Ir/TiO2 Au/TiO2	Catalytic activity decreases in the following order:$Pt/TiO_{(2)}$>$Pd/TiO_{(2)}$>>Rh/TiO_2~ Ir/TiO2>>Au/TiO2.	[55]
Methyl isobutyl ketone	Pt/Zeolites	-Active at 185°C. -Oxidation is affected by: the acidity of the substrate, the Si/Al ratio and the Pt content.	[56]
Toluene	Pd/MgO, Pd/Al2O3, Pd.O,. Pd/SiO2, Pd/SnO2, Pd/WO3 and Pd/ZrO_2	-The acid-base properties of the support influence the active phase-support interaction, which in turn controls the activity of Pd. -Good stability of Pd on ZrO_2. -Pd/ZrO_2 is the most active	[57]
Chlorobenzene	V;O,TIO; V;O, sulphated TiO;	-V_2O_5/TiO_2 less active than sulphated $V_2O_5/TiO_{(2)}$. .The sulphate species improve the acidity of the solid. - The interaction between the sulphate groups and the vanadium species improves the reactivity of the vanadium redox sites.	[58]
Ethylene acetate	M (Fe, Co, Cu, Cr and Mn)/Al2O3Ce0.5Zr0.5O2	-CuM.l2O3Ce0.5Zr0.5O2 and MnMl2O3Ce0.5Zr0.5O2 are the most active: conversion reaches 99% at 245°C. -Good dispersion and reduction of Mn and Cu on	[59]

		$Al2O3Ce0.5Zr0.5O2$.	
Benzene and Acetaldehyde	Cu-Mn/TiO2 Cu-Mn/ZrO_2 Cu-Mn/AhO3	-Catalytic activity is influenced by the nature of the support and its redox properties. It follows the following decreasing order: Cu-Mn/$TiO_{(2)}$> Cu-Mn/$ZrO_{(2)}$> Cu-Mn/Al_2O_3.	[60]
Toluene	Cr-Cu/SiO2 Cr-Cu/SBA-15	-Cr-Cu/SBA-15 is much more active than Cr-Cu/SiO2. - Solid activity is influenced by the nature of the substrate.	[61]

Work reported in the literature has shown that noble metals, in particular palladium and platinum, are the most active in the oxidation of VOCs [45-49]. However, their high cost now limits their use and they are being replaced by transition metals such as manganese (Mn), iron (Fe), copper (Cu), cobalt (Co) and silver (Ag). The latter have also shown significant activity in the oxidation of volatile organic compounds such propene, chlorobenzene, ethylene acetate, benzene and toluene [58-62].

I.2.5.2. Use of silver catalysts in the oxidation of VOCs

Silver is a transition element to the right of palladium in the periodic table (line 5). Its physicochemical characteristics and oxidising properties can be similar to those of Pd. As a result, a great deal of work has focused on the use of Ag-containing catalysts in the oxidation of volatile organic compounds. Indeed, Lj. Kundakovic et al. [62] tested a series of catalysts containing 1%, 2%, 3.5%, 5%, 6.5%, 7.9% or 8% Ag supported on zirconia in the CH4 combustion reaction. They showed that the catalysts are active in CH4 oxidation for T > 300°C and that the catalytic activity is affected by the silver content and the oxidation state and dispersion of the Ag particles.

Z. Guerra-Que et al [63] studied the wet oxidation reaction of methyl tert-butyl ether on Ag/ZrO2-CeO2 solids (with CeO2 = 0 ; 0.5 ; 1 ; 5 ; 10 ; 15 and 20 wt. %). They showed that the addition of cerium favours the formation of Ag2O species and the 15 wt. % CeO2 content leads to the best catalytic activity.

In the combustion of carbon black: K. Schimizu et al [64] explained the catalytic activity of Ag/CeO2 solids by the good dispersion of the Ag2O species on the surface of the support. G. Corro et al [65] showed that the catalytic activity of Ag/SiO2 solids is due to the presence of metallic Ag and silver oxides. K. Yamazaki et al[66] reported that a strong interaction between Ag and Ce allows the formation of oxygenated O_n^{x-} species (O^{2-} being one of them) at the surface of the Ag/CeO2 catalyst which participate in the oxidation of carbon black.

S. Imamura et al [67] noted that ceria, in the Ag/CeO2 catalyst, plays an important role in the oxidation of carbon monoxide (CO) at low temperatures due to its high oxidation-reduction capacity and because it allows good dispersion of the silver species on the catalyst surface.

Z. Qu et al [68] reported that total CO conversion is obtained at T < 65°C over a mesoporous Ag/CeO2 catalyst. Good catalytic activity is also obtained over Ag/SBA-15 catalysts and is affected by the nature of the gas used for catalyst pre-treatment and by the amount of silver [69].

Q. Ye et al [70] have shown that 5% wt Ag is the optimum amount silver for obtaining better catalytic activity in benzene and CO oxidation on x% Ag/nanoMnO2 type catalysts (1%<x<10%).

M. Skaf et al [71] studied the effect of the preparation method (impregnation (Imp) and precipitation deposition (DP)) on the catalytic power of the Ag/CeO2 solid in the oxidation of propene, carbon monoxide and carbon black. They reported that the solid prepared by Imp is more active than that prepared by DP and this is due to the presence of the redox couples Ag^{2+}/Ag^{+} , Ag^{2+}/Ag^{0}, $Ag^{+}/Ag(0)$ in the first catalyst only.

I.2. 5.3. Oxidation mechanisms of VOCs

The total catalytic oxidation of a simple VOC (containing C and H only) takes place according to the following balance equation:

$$C_mH_n + (m + n/4)\ O_2 \rightarrow n/2\ H_2O + m\ CO_2$$

According to the literature, the first stage of this reaction corresponds to the strong adsorption of the VOC on the active centre of an anionic oxygen atom of the catalyst and the formation of an activated complex which then reacts to give the products of the reaction. These desorb from the surface of the catalyst and the lost anionic oxygen is replaced by another gaseous oxygen thanks to the existing interaction between the support and gaseous O2 [72].

The reaction mechanisms reported in the literature for the oxidation of VOCs are : Eley Rideal, Langmuir Hinshelwood and Mars-Van Krevelen. In these mechanisms, a certain mobility of the surface oxygen and a redox cycle taking place at the surface of the catalyst between surface O2 (chemisorbed or belonging to the lattice) and a chemisorbed or gas-phase reactant are necessary [72].

In the Eley Rideal (ER) mechanism, a direct reaction takes place between a reagent existing in the gas phase and an oxygen adsorbed on the surface of the catalyst. Aranzabal et al [73] used this mechanism to explain the total oxidation of trichloroethylene (TCE) on the Pd/Al_2O_3 solid.

In the Langmuir Hinshelwood (LH) mechanism, the reactant adsorbs to the surface of the catalyst and then reacts with the adsorbed oxygen to give the products, which are then desorbed from the surface of the solid. GotzVeser et al [74] explained the oxidation of several alkanes (methane, ethane, propene and isobutane) over noble metal catalysts (Pt, Pd, Rh, Ir) by this type of mechanism.

The Mars-Van Krevelen (MVK) mechanism is the most widely reported for the oxidation of VOCs on metal oxides. In this mechanism, the reactant is oxidised by an oxygen on the oxide support, which is reduced and a vacant site is created on the surface. This site is again occupied by an oxygen thanks to a reaction between the support and gaseous oxygen. P.Mars, D.W.Van Krevelen [75] have reported that the oxidation of naphthalene on V2O5 occurs according to this mechanism and following these two stages:

HC	+	KO	→ HCO	+	K
Réactif		**Cat.OX**	**Produit**		**Cat.réd**

K	+	$\frac{1}{2} O_2$	→	KO
Reagent		**Cat.OX**	**Product**	**Cat.red**

> *In the first stage,* the reactant (hydrocarbon HC) co-fed with oxygen O2 (diluted or not) is transformed into an oxygenated product "HCO" (generally with the production of water) thanks to the oxygen O^{2-} of the catalyst in its oxidised form (KO).

> *In the second stage,* the reduced catalyst (K) is reoxidised to KO by oxygen.

The Mars-Van Krevelen mechanism [47,64,69,76-79] is widely used to explain the catalytic oxidation of VOCs (such as propene, carbon monoxide, carbon black and certain hydrocarbons and C_3 oxygenates) over catalysts such as Mn_3O_4, CeO_2, CuO, Ag/CeO2 and Fe_2O_3.

In addition to its redox properties and the high mobility of oxygen, the acidity of a catalyst also influences its activity in the oxidation of VOCs [73-77]. According to Popovas et al [78], the acid-base properties of the catalyst affect the sorption of organic molecules and consequently their conversion in oxidation reactions. Papaefthimiou et al [79] obtained better oxidation of ethyl acetate at low temperatures over Pt/TiO2 (w $^{6+}$) than over Pt/TiO2. They attributed this to the acidity, which is greater for Pt/TiO2 (w $^{6+}$). Furthermore Okumura et al [80] noted that the acid-base properties of the support (Al2O3, Nb2O5, SnO2, WO3 and ZnO2) influence the interaction between the active phase and the support and affect the catalytic activity of palladium.

I. 3 Catalytic oxidation of toluene

I. 3. 1 Toluene

Toluene is an aromatic hydrocarbon derived from benzene, with the chemical formula C7H8. At room temperature, it is a colourless, highly volatile liquid with an extremely strong, benzene-like odour [21, 83-84]. The physicochemical properties of this compound are shown in Table 7.

Table 7. Physicochemical properties of toluene.

Name	Methylbenzene or Phenylmethane
Chemical formula	C7H8
Physical condition	Liquid
Molar mass	92.14 g/mol
Density	867 kg/m^3
Boiling point	110,6 °C
Melting point	-95°C
Auto-ignition temperature	535°C
Vapour pressure	3 kPa at 20 °C and 3.8 kPa at 25 °C
Density	0,866
Vapour density	3,14
Stability	Stable

I.3.2 Harmful effects of toluene

In humans, the main route of exposure to toluene is inhalation, where absorption is around 50% (proportional to lung ventilation). Absorption by mouth is 100%. Table 8 summarises the main effects of brief and prolonged exposure to toluene [85-86].

Table 8. Harmful effects of toluene on health.

Nature of exposure to toluene	***Harmful effects on human health***
Brief (acute risk)	*Ingestion of toluene:* Abdominal pain, nausea, vomiting, diarrhoea, disturbed consciousness (even coma). *Inhalation of toluene:* cough, fever, pneumonitis, dyspnoea, euphoria, hallucination, fatigue, dizziness, muscle weakness, insomnia, mental confusion, coordination disorders, reduced heart rate and respiratory arrest (depending on the concentration and duration of exposure to toluene vapours). *Contact with toluene:* Eye irritation and irritant dermatitis.
Prolonged (chronic risk)	Impaired memory and concentration, insomnia, reduced intellectual performance (depending on the extent of exposure). Chronic irritant dermatitis: linked to the drying and degreasing action of toluene on the skin.

I.3.3 Catalysts used in toluene oxidation

Several types of catalyst have been tested in the oxidation of toluene (C_7H_8). H. J. Kim et al [87] showed that the Mn-Ce/Al_2O_3 solid is characterised by good oxygen mobility and better catalytic activity in the oxidation of C_7H_8 compared with the Mn/Al_2O_3 and Ce/Al_2O_3 catalysts. While S. C. Kim et al [88] noted that the activity of Mn oxides in the oxidation reaction of benzene and toluene decreases in the following order: Mn_3O_4 > Mn_2O_3> MnO_2.

K. Jiratovaet al [89] reported that the addition of 1% wt. potassium (K) to the Co-Mn-Al catalytic system modifies its acid-base properties and improves its activity in the total oxidation of toluene.

M. Hosseini et al [90-91] studied the effect of the composition of the core and shell of a Pd-Au/TiO_2 system on its catalytic power in toluene oxidation. They concluded that a system with an Au-rich core is more active than a system with a Pd-rich core.

T. Barakat et al [72] studied the effect of the nature of the Pd precursor on the catalytic activity of the Pd/TiO_2 solid in toluene oxidation and showed that Pd(NO_3) leads to the most active solid.

J. G. Guo et al [92] have shown that the CuO/zeolite material containing ceria is more active in the oxidation of several VOCs (ethanol, acetone, toluene and benzene) than that prepared without ceria.

X. Liang et al [93] reported that the amount of iron influences the total toluene conversion on Fe-Ti-PILC catalysts. The best catalytic activity is obtained for an iron content of 3.75%. In addition, the catalysts exhibit high stability in the presence of 50,000 ppm water .

I.3.4 Use of Ag catalysts in toluene oxidation

Despite the fact that silver-based catalysts (Ag/SiO_2, Ag/CeO_2, Ag/ZrO_2, Ag/SBA-15, etc.) have shown good catalytic performances in the oxidation of VOCs (carbon monoxide, benzene, methane, carbon black, propene), they are, according to the bibliography, little used in the toluene oxidation reaction [94-97] . For example, M. Xue et al [94] showed that the Ag/V_2O_5 catalyst is active in toluene oxidation and leads to benzaldehyde and benzoic acid products. Ge et al [95] noted that the addition of silver to the V_2O_5/TiO_2 solid increases its catalytic activity and improves its selectivity to benzaldehyde and benzoic acid. Z. Qu et al [96] found that the Ag-Mn/SBA-15 catalyst was more active than those containing only Ag or Mn. It leads to total oxidation of toluene to CO_2 at 260°C. J. Li et al [97] revealed that the structure and morphology of MnO_2 affects the catalytic activity of the Ag/MnO_2 solid in toluene oxidation. This activity decreases in the following order: Ag/MnO_2 nanowires> Ag/MnO_2 nanorods> Ag/MnO_2 nantube

X. Zhang et al [98] studied the effect of Ag content on the catalytic properties of the Ag/UiO-66 solid. They showed that solid 10Ag-U is the most active between 295 and 400°C due to the presence of Ag^0 on its surface. It oxidises toluene to benzaldehyde and benzoic acid.

M. Lu [99] reported that the sample 2%FeOx-1%AgOy/SBA-15 is more active in the toluene oxidation reaction with non-thermal plasma (NTP) assistance than the solids SBA-15 and 3%FeOx/SBA-15 because the presence of Ag improves the dispersion and reduction of FeOx. Y.Qin et al [100] studied the effect of pretreatment conditions (O_2 at 500 °C (O500), H_2 at 500 °C (H500), O_2 at 500 °C, and H_2 at 300 °C (O500-H300)), on the activity of Ag(x)/SBA-15 catalysts with x = 6, 8, 10, 12, 16, and 25%. They showed that the Ag (12)/SBA-15 (O500-H300) catalyst exhibits excellent catalytic activity, which is due to the presence of Ag nanoparticles well dispersed on the surface that are able to activate O_2 easily.

I. 3.5 Preparation methods for catalysts used in toluene oxidation

The most commonly used methods for preparing catalysts for oxidation reactions are impregnation, co-precipitation and precipitation [101-107].

I.3.5.1 Impregnation

In this method, the substrate is brought into contact with a metal precursor solution. It is said to be dry if the volume of the solution is equal to the pore volume of the substrate, whereas it is said to be in excess of solvent if the volume of the solution is greater than the pore volume of the substrate. The solvent is then removed by drying, which promotes the formation of precursor crystals inside the pores of the substrate. Depending on whether there are weak or strong interactions between the metal precursors and the substrate, there are two types of impregnation: immersion and ion exchange. Cation-exchange zeolites are a good example of the latter.

Immersion (or wetting) simply involves bringing the substrate into contact with the solution of the metal precursor. This solution is distributed into the pores of the substrate by capillary action. Stirring at room temperature ensures satisfactory homogeneity. Finally, drying (which is generally carried out by simple evaporation) leads to the formation of precursor crystals inside the pores of the substrate.

1.3.5.2. Co-precipitation

This method consists of simultaneous precipitation of the active phase and the support in a solution by the action of a precipitating agent [108]. The solid obtained is then dried and activated by a suitable heat treatment. Using this technique, metal salts of different metals can be deposited simultaneously on the support.

It should be noted that metal is rarely used in a solid state because its specific surface area is too small. This is why it is often deposited on a support, at a rate of a few per cent by mass, increase its active surface.

1.3.5.3. The rush

In this case, a solution of a metal salt is brought into contact with a support. The mixture is then precipitated by adding a base (such as KOH) and kept under stirring with or without heating. Finally, the solid is recovered and washed with water [109].

It has been reported in the literature that the method of preparing catalysts has a major influence on their physicochemical and catalytic properties. In this work we have used the sol-gel method for the preparation of catalyst supports because it offers many advantages, namely :

- Obtaining homogeneous materials of high purity (due to the dispersion of precursors in solution).
- Synthesis of very fine powders
- Better control of texture, composition and structure than with conventional methods.
- The development of specific large-surface materials
- A great mastery of doping.

1.3.5.4. Sol-gel process

1. 4. 1 Definitions

Sol-Gel is defined as a suspension of small colloidal particles in a liquid. Following the association and aggregation of colloidal particles by chemical reactions, it can evolve into a network of infinite viscosity called a "gel". The aggregates of this gel trap the solvent in a three-dimensional network [110]. Drying of the gel (elimination of the solvent) causes the volume to shrink and can lead to two types of material:

> A xerogel if dried in an oven under atmospheric pressure.

> An aerogel if drying takes place in an autoclave (with supercritical solvent conditions).

The various stages involved in sol-gel synthesis are shown in Figure 1.

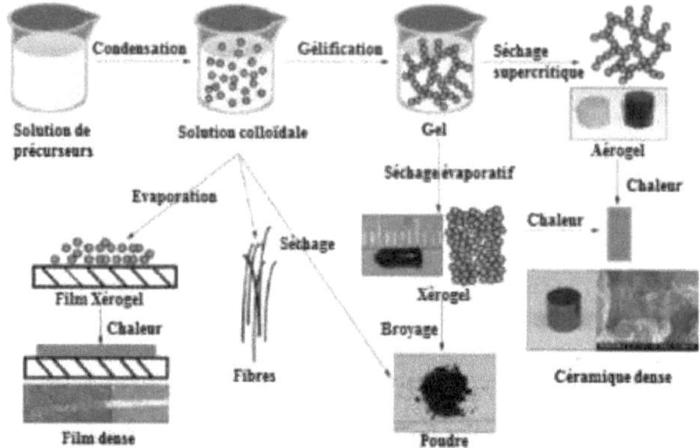

Figure 1. The different stages in sol-gel synthesis.

The sol-gel method involves two routes: the inorganic route, where the precursors are metal salts dissolved in an aqueous solution, and the organic route, where the precursors are metal alkoxides mixed in an organic solution with water.

I. 4.2 Metal alkoxides

Metal alkoxides are compounds of the general formula M(OR)n, where OR is an alkoxy group and M is an n-valent metal. J. Livage et al [111] have reported that transition metal alkoxides ($Zr(OEt)_4$, $Ti(oEt)_4$, $Nb(OEt)_5$, $Ta(OEt)_5$, $VO(OEt)_5$ and $W(OEt)_6$) are extremely reactive with water due to the presence of an electronegative alkoxy group that exposes the metal atom to nucleophilic attack [112].

I. 4.3 Hydrolysis and condensation reactions

The mechanism of the sol gel method is based on two types of reaction: hydrolysis and condensation, which are described in detail in the following paragraphs.

1. 4. 3. a) Hydrolysis reactions

This reaction leads to the formation of "sol" and occurs between the alkoxide and the water according to the balance equation (1)

$$M\text{-}(OR)_n + x\,H_2O \;\hat{}\; M(OH)_x(OR)_{n-x} + x\,ROH \quad (1)$$

Ж The hydrolysis steps are:

- Nucleophilic attack of the metal by the water molecule (a^b)
- Proton transfer from water to the (OR) group (b^c)
- Elimination of an alcohol molecule (c^d)

Ж The mechanism alkoxide hydrolysis is as follows:

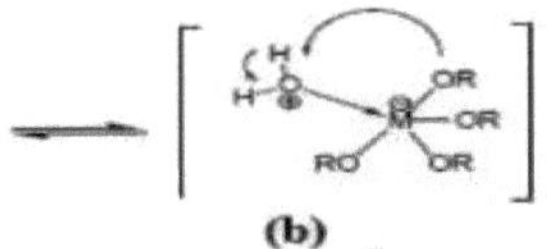

(b)

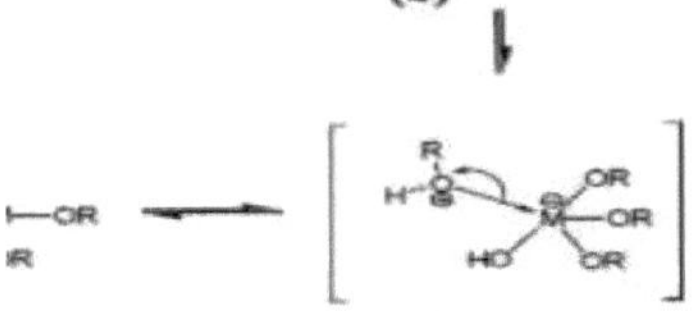

(c)

I. 4. 3. b) Condensation reactions

Condensation begins as soon as the active groups M(OR)n-x-OHx are formed. These react with each other or with an alkoxide molecule M(OR)n to form a gel. During condensation, different mechanisms can take place: alkoxolation, oxolation, alcoholation and olation.

is Alkoxolation: Formation of an "M-O-M" oxo bridge and elimination of a ROH molecule according to equation (2).

M-OH + RO-M →M-OM + ROH (2)

Ж The alkoxidation mechanism for alkoxides is as follows:

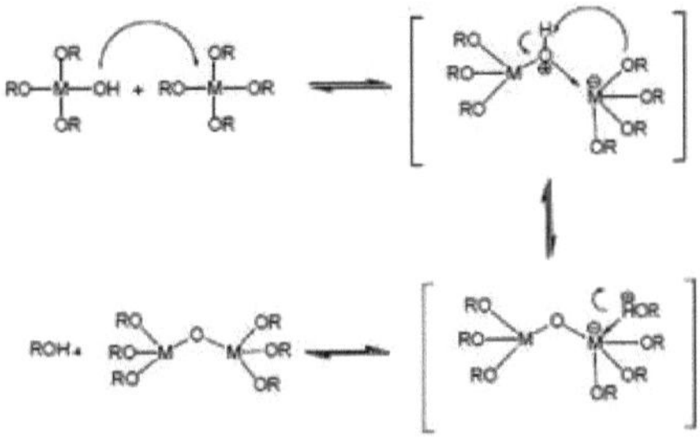

s3 Oxolation: Formation of an oxo bridge by a reaction between two M-OH molecules with the elimination of a molecule of water according to equation (3).

M-OH + HO-M →M-O-M + H_2O (3)

The oxidation mechanism for alkoxides is as follows:

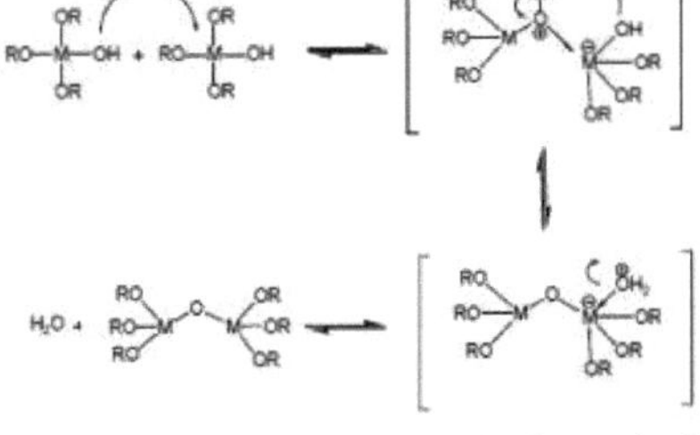

Alcoholation: In this case the terminal ligands (OR), linked to a single metal centre, form a bridge between the metal centres according to the following mechanism:

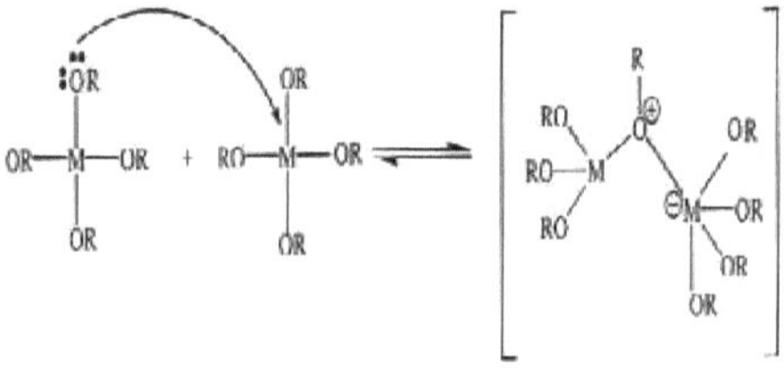

O Olation: formation of an M-OH-M hydroxo bridge with elimination of an ROH or H_2O molecule by the following mechanism:

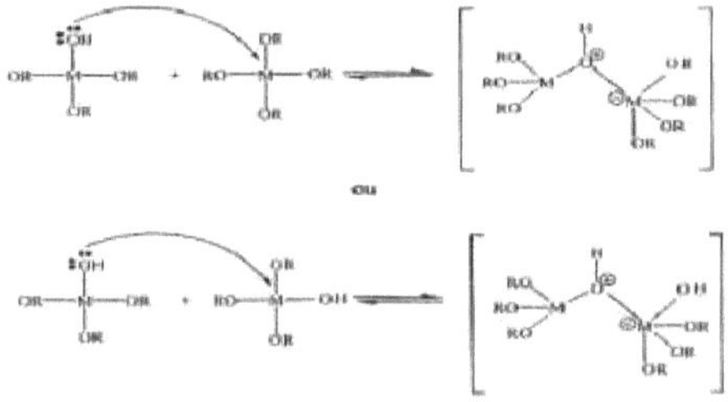

ou

I. 4.4 Parameters influencing Sol-Gel preparation

Several preparation factors can modify the final state of the material obtained, such as: internal parameters (nature of the metal and the OR group, structure of the precursor, etc.) and external parameters (hydrolysis and complexation ratios, solvent, pH, temperature, alkoxide concentration, etc.).

I.4. 4. a) Effect of the metal atom and the alkyl group

The reactivity of metal alkoxides depends on the electrophilic nature of the metal and its degree of oxidation (N-Z) [111]. Generally, the degree of oxidation Z of the metal is smaller than its coordination number N [113], so the metal tends to increase its coordination number by forming oligomers [111]. For example, the oxidation state of aluminium (+3) is lower than its coordination number (+6), so aluminium alcoholates are very reactive and are generally found in the form of oligomers, whereas silicon alcoholates are less reactive since for silicon the unsaturation number (N-Z) is zero [111]. In addition, the degree of oligomerisation of metal alkoxides decreases when the branching of the alkyl group increases because of the steric effect of this group [114].

I. 4. 4. b) Effect of solvent

Metal alkoxides are not soluble in water but are soluble in alcohols. Note that when the alkoxide is solubilised in an alcohol with the same alkoxy OR group, the molecular complexity is reduced and the reactivity with respect to hydrolysis increases.

When the alcohol used does not have the same OR group as the alkoxide, an exchange reaction can occur according to following equation:

$$M(OR)_n + x\,R'OH \wedge M(OR)_{n-x} + x\,ROH$$

Exchange is easier when the steric hindrance of the alkoxy group is low. It follows the following order: MeO > EtO > PriO > ButO and it affects the gelation time and the physicochemical properties of the prepared sample. Indeed, C.Sanchez et al [112] have shown that gels prepared from $Si(OMe)_4/MeOH$ and $Si(OMe)_4/EtOH$ solutions are not characterised by the same gelation time and lead to powders with different specific surface areas; 305 m^2/g and

169 m^2/g, respectively.

1. 4. 4. c) Effect of hydrolysis ratio (h)

Several studies have shown that the hydrolysis ratio (h= nH2O/nM(OR)n) has a major influence on the kinetics of the hydrolysis and condensation reactions. C. Sanchez et al [112] have shown that the gelation time can be fixed by means of this parameter and three domains are considered:

- ***If h < 1*:** Hydrolysis is controlled by the local absence of water and condensation occurs by alkoxolation and alkolation. Gelling or precipitation cannot occur due to the lack of water.
- ***If 1 <h <n*:** Rapid hydrolysis takes place and condensation can occur by alkoxolation and oxolation. If the partial charge of the OR group is positive, alkoxolation is thermodynamically favoured and if the maximum co-ordination of the metal is zero (N-Z = 0), oxolation becomes competitive. Note that in the case of transition metal alkoxides, olation is favoured because of the favourable charge conditions: (δ (OH) << 0 or δ (M) >> 0 and N-Z >> 0).
- **If h > n:** Hydrolysis is favoured by excess water and condensation produces gels or gelatinous precipitates, probably by the olation mechanism (not oxolation).

I.4. 4. d) Effect of complexing agents

To avoid rapid precipitation and control the properties of the solids prepared (size, morphology, etc.), complexing agents such as nucleophilic molecules (XOH) are added. These agents make it possible obtain new molecular precursors that are more stable than the initial precursors, according to the following reaction:

$$M (OR) + x XOH \wedge M (OR)_{n-x} (OX)_x + xROH$$

Other complexing agents used include inorganic acids, carboxylic acids and β-diketones [110]. The effects of several metal-e-diketonates on different alkoxides are reported in the literature. The work shows that the enolic form of β-diketonates is stabilised by binding to the metal resulting in bidentate complexes that are stable and less sensitive to water [113-116]. Acetylacetonate is a strong complexing agent [110-112] which is used for several metal alkoxides such as: $Al(s\text{-}OBu)_3$ [115],$W(OEt)_6$ [118], $Zr(i\text{-}OPr)_4$[115], $Ti(i\text{-}Opr)_4$ [119] and $Ti(n\text{-}OBu)_4$ [120].

I. 4. 4. e) Effect of pH

The pH also influences the physico-chemical properties of the prepared sample. For example, R. Caruso et al [121] studied the effect of pH during the preparation of ZrO_2 powders. They reported that the solution obtained during the hydrolysis of zirconium n-propoxide is yellowish at pH = 0.5 and whitish at pH = 5.

I. 4. 4. f) Effect of temperature

An increase in temperature activates the hydrolysis and condensation processes, so for highly reactive precursors such as transition metal alkoxides (Nb, Zr, etc.), the temperature must be lowered to reduce the hydrolysis and condensation rates [112]. For $Si(OR)_4$, which is not very reactive, the temperature must be increased.

I. 4. 4. g) Effect of drying

Drying the gel in the oven leads to an increase in the size of the crystallites and a decrease in the specific surface area and certain pores of the xerogel solid. On the other hand, drying under supercritical solvent conditions results in an "aerogel" solid characterised by large porosity, low density and good thermal stability.

1.5. Conclusion

The bibliographical study presented in this chapter has shown that toluene is a dangerous volatile organic compound which has effects on humans and the environment, hence the need to eliminate this compound. Catalytic oxidation is an effective technique for treating toluene. Several catalysts, prepared by different methods, are being tested in this reaction. Silver-based catalysts have also shown good catalytic activity in this reaction, so our work involved the preparation (using the impregnation method) and characterisation of new Ag/Ce-ZrO2 and Ag/Fe-ZrO2 catalysts (containing varying amounts of Ag) for the total oxidation of toluene. The Ce- ZrO2 and Fe-ZrO2 supports were prepared by the sol gel method with different Ce and Fe contents.

1. 6. References

1. Z. Qu, Y. Bu, Y. Qin, Y. Wang, Q. Fu, *The improved reactivity of manganese catalysts by Ag in catalytic oxidation of toluene,* Appi. Catal. B: Environ. 132- 133 **(2013)** 353- 362.
2. S. Todorova, A. Naydenov, H. Kolev, J.P. Holgado, G. Ivanov, G. Kadinov, A. Caballero, *Mechanism of complete n-hexane oxidation on silica supported cobalt and manganese catalysts*, Applied Catalysis A: General 413- 414 **(2012)** 43- 51
3. M. S. Kamal, S. A. Razzak , M. M. Hossain, *Catalytic oxidation of volatile organic compounds (VOCs) e A review*, Atmos. Environ 140 **(2016)** 117e134
4. C. Yang, G. Miao, Y. Pi, Q. Xia, J. Wu, Z. Li, J. Xiao, *Abatement of various types of VOCs by adsorption/catalytic oxidation: A review,* Chem. Eng. J. 370 **(2019)** 11281153
5. R.E. Hayes, S. Kolaczkowski, *Introduction to Catalytic Combustion, Gordon and Breach Science Publisher*, Amsterdam, **(1997).**
6. M.C. Alvarez-Galvan, V.A. de la Pe~naO'Shea, J.L.G. Fierro, P.L. Arias, *Alumina-supportedmanganese- and manganese-palladium oxidecatalysts for VOCs combustion*, Catal. Commun. 4 **(2003)** 223-228.
7. I.Suh, R.Zhang, L.T. Molina and M.J. Molina, *Oxidation Mechanism of Aromatic Peroxy and Bicyclic Radicals from OH-Toluene Reactions*, J. Am. Chem. Soc. 125 **(2003)** 12655-12665.
8. J. Dewulf, H. Van Langenhove, G. Wittmann, *Analysis of volatile organic compounds using gas chromoatography*, TrAC Trends Anal. Chem. 21 **(2002)** 637-646.
9. L.F. Liotta, *Catalytic oxidation of volatile organic compounds on supported noble metals,* Appi. Catal. B: Environ.100 **(2010)** 403-412.
10. S. Scirè, S. Minicò, C. Crisafulli, C. Satriano , A.Pistone, *Catalytic combustion of voltalis organic compounds on gold/ceriu oxide catalysts*, Appl. Catal. B: Environ. 40 **(2003)** 43-49.
11. R. Balzer, L. F. D. Probst, V. Drago, W. H. Schreiner and H. V. Fajardo, *Catalytic oxidation of volatile organic compounds (n-hexane, Benzene, Toluene, o-Xylene) promoted by cobalt catalysts supported on Y-Al_2O_3-CeO2*, Braz. J. Chem. Eng. 31 **(2014)** 757-769.
12. H.L. Tidahy, S. Siffert, J.-F.Lamonier, E.A. Zhilinskaya a, A. Aboukais,Z.-Y. Yuan , A. Vantomme , B.-L. Su , X. Canet , G. De Weireld, M. Frère , T.B. N'Guyen, J.- M. Giraudon , G. Leclercq, New Pd/hierarchical macro-mesoporous ZrO2, TiO2 and ZrO2-TiO2 catalysts for VOCs totalM. Giraudon , G. Leclercq, *New Pd/hierarchical macro-mesoporous ZrO2, TiO2 and ZrO2-TiO2 catalysts for VOCs total oxidation*, Appl. Catal. A: Gen. 310 **(2006)** 61-69.
13. S. Todorova, A. Naydenov, H. Kolev, J.P. Holgado, G. Ivanov, G. Kadinov, A. Caballero, *Mechanism of complete n-hexane oxidation on silica supported cobalt and manganese catalysts*, Appl.Catal.A: Gen. 413- 414 **(2012)** 43- 51.
14. T.Miki and Y. Tai, *Catalytic oxidation of toluene over Fe2O3/Al2O3 catalyst*, Mater. Sci. Forum, 695 **(2011)** 101-104.
15. J. Carpentier, J.F. Lamonier, S. Siffert, E.A. Zhilinskaya, A. Aboukais, *Characterisation*

of Mg/Al hydrotalcite with interlayer palladium complex for catalytic oxidation of toluene, Appl. Catal.A: Gen, 234 **(2002)** 91-101.
16. J. L. Kim, L. Elfman, Y. Mi, G. Wieslander, G. Smedje, D. Norbck, *Indoor moulds, bacteria, microbial volatile organic compounds and plasticizers in schools - associations with asthma and respiratory symptoms in pupils*, Indoor Air 17 **(2007)** 153-163.
17. F.H. Sennhauser, C. Braun-Fahrlander and J. H. Wildhaber. *The burden of asthma in children*: a European perspective, Paediatr. Respir. Rev. 6 **(2005)** 2-7.
18. M. Phillips, K.Gleeson, J. Michael B Hughes, J. Greenberg, R.N Cataneo, L.Baker, W P. McVay,*Volatile organic compounds in breath as markers of lung cancer: a cross-sectional study,* **(1999)** https://doi.org/10.1016/S0140-6736(98)07552-7.
19. H. Guo, S.C. Lee, L.Y. Chan, and W.M. Li, *Risk assessment of exposure to volatile organic compounds in different indoor environments,* Environ. Res. 94 **(2004)** 57-66.
20. P. Le Cloirec, *Les composés organiques volatils (COV) dans l'environnement*, Technique et documentation Lavoisier, Paris **(1998).**
21. R .Atkinson, *Atmospheric chemistry of VOCs and NOx*, Atmos. Environ. 34 **(2000)** 2063-2101.
22. M.J.Molina, *The Role of Chlorine in Stratospheric Chemistry*, Pure and Appl.Chem. 68 **(1996)** 1749-1756.
23. *Volatile Organic Compounds of the European Ozone Directive*, October **(2005).**
24. J. Li, J. Zhang, Z. Lei, and B. Chen, *Pd-Co Coating onto Cordierite Monoliths as Structured Catalysts for Methane Catalytic Combustion*, Energy Fuels 26 **(2012)** 443450.
25. M. Herzog, *Characterisation and control of odours and VOC in the progress industries,* ed.Vigneron, H., Chaouki - Elsevier Science B.V **(1994).**
26. Z. Jiang, J. Yu, J. Cheng, T. Xiao, M.O. Jones, Z. Hao, P.P. Edwards, *Catalytic combustion of methane over mixed oxides derived from Co-Mg/Al ternary hydrotalcites,*Fuel Process. Technol. 91 **(2010)** 97-102.
27. A.C. Gluhoi, B.E. Nieuwenhuys, *Catalytic oxidation of saturated hydrocarbons on multicomponent Au/Al2O3 catalysts: Effect of various promoters*, Catal.Today 119 **(2007)** 305-310.
28. P. Mocho, PhD thesis, *Adsorption of volatile organic compounds on activated carbon. In situ regeneration of carbon by electromagnetic induction heating,* **(1994).**
29. X. Wang, R. Daniels, R. W. Baker. *Recovery of VOCs from high-volume, low -VOC-concentration air streams,* AICHE J. 47 **(2001)** 1094-1100.
30. P. L. Cloirec, *Les composés organiques volatils (COV) dans l'environnement,* Tec & doc-Lavoisier **(1998).**
31. F. Cotte. Doctoral thesis, *Absorption in packed columns and spray towers. Application au traitement de composés organiques volatils*, **(1996).**
32. S. Cimino, S. Colonna, S. D. Rossi, M. Faticanti, L. Lisi, I. Pettiti, P. Porta, *Methane Combustion and CO Oxidation on Zirconia-Supported La, Mn Oxides and LaMnO3 Perovskite*, J. Catal. 205 **(2002)** 309-317.
33. G. Laugel, J. Arichi, P. Bernhardt, M. Molière, A. Kiennemann, F. Garin, B. Louis, *Preparation and characterisation of metaloxides supported on SBA-15 as methane combustion catalysts*, Comptes Rendus Chim. 12 **(2009)** 731-739.
34. P. Forzatti, *Environmental catalysis for stationary applications*. Catal. Today 62 **(2000)** 51-65.
35. L. A. Isupova, G.M. Alikina, O.I. Snegurenko, V. A. Sadykov, S. V. Tsybulya, *Monolith*

honeycomb mixed oxide catalysts for methane oxidation, Appl. Catal. B Environ. 21 **(1999)** 171-181.
36. M.C. Alvarez-Galvan, V.A. de la Pe~naO'Shea, J.L.G. Fierro, P.L. Arias, *Alumina-supported manganese- and manganese-palladium oxide catalysts for VOCs* combustion, Catal.Commun. 4 **(2003)** 223-228.
37. M. F. Finol, J. Rooke, B. L. Su., M. Trentesaux, J. M. Giraudon, J. F. Lamonier, *Additional effects of Pt and Nb on hierarchically porous titania in the catalytic removal of n-butanol*, Catal. Today, 192 **(2012)** 154-159.
38. M. Idrissi, J. F. Lamonier, D. Chlala, J. M.Giraudon , M. Chaouch, Y. Miyah, F. Zerrouq, *Treatment of toluene present in trace amounts in air, in the presence of Cu-catalysts Bentonite*, J. Mater. Environ. Sci. 5 (S1) **(2014)** 2303-2308.
39. H.L. Tidahy, S. Siffert , F. Wyrwalski, J.-F. Lamonier, A. Aboukais, *Catalytic activity of copper and palladium based catalysts for toluene total oxidation*, Catal. Today, 119 **(2007)** 317-320.
40. A. Janbey, W. Clark, E. Noordally, S. Grimes, S. Tahir, *Noble metal catalysts for methane removal*, Chemosphere 52 **(2003)** 1041-1046.
41. J. Li, J. Zhang, Z. Lei, B. Chen, *Pd-Co Coating onto Cordierite Monoliths as Structured Catalysts for Methane Catalytic Combustion*, Energy. Fuels 26 **(2012)** 443450.
42. K. Kohse-Höinghaus, P. Oßwald, T. A. Cooll, T. Kasper, N. Hansen, F. Qi, C. K. Westbrook, P. R. Westmoreland, *Biofuel Combustion Chemistry: From Ethanol to Biodiesel*, Angew. Chemie Int. Ed. 49 **(2010)** 3572-3597.
43. S.Sebti, H.Boukhal, N.Hanafi and S. Boulaajaj, *Catalyse de la r6action de Michael par le fluorure de potassium support6 sur le phosphate naturel*, TetrahedronLett. 40 **(1999)** 6207-6209.
44. J.F. Le Page, *Catalyse de Contact: Conception, Préparation et Mise En Oeuvre Des Catalyseurs Industriels*, **(1978).**
45. T.Garcia, S. Agouram, J. F. Sanchez-Royo, R. Murillo, A. M. Mastral, A. Aranda, I.Vazquez, A. Dejoz, B. Solsona, *Deep oxidation of volatile organic compounds using ordered cobalt oxides prepared by a nanocasting route,* Appl. Catal.A: Gen. 386 **(2010)** 16-27.
46. Q. Dai, X. Wang and G. Lu, *Etude des propriétés catalytiques de mélanges oxyde/ Faujasite pour le traitement de compsés organiques volatils (COVs) : caractérisation-mobilité de l'oxygène-oxydation*, Catal. Com. 8 **(2007)** 1645-1646.
47. M. Baldi, V. S. Escribano, J. Manuel Gallardo Amores, F. Milella, G. Busca, *Characterization of manganese and iron oxides as combustion catalysts for propane and propene*, Appl. Catal. B: Environ. 17 **(1998)** L175-L182.
48. B. de Rivas, R. López-Fonseca, C. Jiménez-González, J. I. Gutiérrez-Ortiz, *Synthesis, characterisation and catalytic performance of nanocrystalline Co3O4 for gas-phase chlorinated VOC abatement*, J. Catal. 281 **(2011)** 88-97.
49. E. Finocchio, G. Busca and V. Lorenzelli, *FTIR Studies on the Selective Oxidation and Combustion of Light Hydrocarbons at Metal Oxide Surfaces*, J. Chem.Soc. Faraday Trans, 90(21) **(1994)** 3347-3356.
50. C. Hu, Q. Zhu, Z. Jiang, Y. Zhang, Y. Wang, *Preparation and formation mechanism of mesoporous CuO-CeO2 mixed oxides with excellent catalytic performance for removal of VOCs*, Microporous. Mesoporous Mater. 113 **(2008)** 427-434.
51. M.F.M. Zwinkels, S.G. Järäs, P.G. Menon, T.A. Griffin, *Catalytic Materials for High-*

*Temperature Combustion,*Catal. Rev. 35 **(1993)** 319-358.

52. R. S. Monteiro, L. C. Dieguez, M. Schmal, *The role of Pd precursors in the oxidation of carbon monoxide over Pd/Al2O3 and Pd/CeO2/Al2O3 catalysts*, Catal. Today 65 **(2001)** 77-89.

53. C. A. Lin, J. C. S.Wu, J. W. Pan and C. T. Yeh, *Characterization of Boron-Nitride-Supported Pt Catalysts for the Deep Oxidation of Benzene*, J. Catal. 210 **(2002)** 39-45.

54. J. Tsou, L. Pinard, P. Magnoux, J.L. Figueiredo, M. Guisnet, *Catalytic oxidation of volatile organic compounds (VOCs) Oxidation of o-xylene over Pt/HBEA catalysts*, Appl. Catal. B: Environ. 46 **(2003)** 371-379.

55. V. P. Santos, S. A.C. Carabineir, P. B. Tavares, M. F.R. Pereira, J. J.M. Orfao, J. L. Figueiredo, *Oxidation of CO, ethanol and toluene over TiO2 supported noble metal catalysts,* Appi. Catal. B: Environ. 99 **(2010)** 198-205.

56. J. Tsou, P. Magnoux, M. Guisnet, J.J.M. Orfao, J.L. Figueiredo, *Catalytic oxidation of volatile organic compounds Oxidation of methyl-isobutyl-ketone over Pt/zeolite catalysts,* Appl. Catal. B: Environ. 57 **(2005)** 117-123.

57. K. Okumura, T. Kobayashi. H. Tanaka, M. Niwa, *Toluene combustion over palladium supported on various metal oxide supports*, Appl. Catal. B: Environ. 44 **(2003)** 325331.

58. C. Gannoun, R. Delaigle, P. Eloy, D. P. Debecker, A.Ghorbel, E.M. Gaigneaux, *Solgel derived V2O5-TO2 mesoporous materials as catalysts for the total oxidation of chlorobenzene,* Catal. Commun. 15 **(2011)** 1-5.

59. Y. S. Hua, S. Mei, G. M. Chu, W.J. Li, Y. S. Hui, C. H. Yan, C.Y. Qiang, *Catalytic Combustion of Ethyl Acetate over Al2O3Ce0.5Zr0.5O2 Supported Metal Oxide Catalysts,* Acta Phys. Chem. Sin. 4 **(2008)** 364-368.

60. P. Doggali, Y. Teraoka, P. Mungse, I. K. Shah, S. Rayalu, N. Labhsetwar, *Combustion of volatile organic compounds over Cu-Mn based mixed oxide type catalysts supported on mesoporous Al2O3, TiO2 and ZrO2*, J. Mol. Catal. A: Chemical 358 **(2012)** 23- 30.

61. M. Popova, Á. Szegedi, Z. Cherkezova-Zheleva, A. Dimitrovac, I. Mitov, *Toluene oxidation on chromium- and copper-modified SiO2 and SBA-15*, Appl. Catal. A: Gen. 381 **(2010)** 26-35.

62. L. Kundakovic, M. Flytzani-Stephanopoulos, *Deep oxidation of methane over zirconia supported Ag catalysts,* Appl. Catal. A: Gen. 183 **(1999)** 35-51.

63. Z. Guerra-Que, G. Torres-Torres, H. P'erez-Vidal, I. Cuauht'emoc-L'opez, A. Espinosa de losMonteros, Jorge N. Beltramini and D. M. Fr'ias-M'arquez, *Silver nanoparticles supported on zirconia-ceria for the catalytic wet air oxidation of methyl tert-butyl ether*, RSC Adv.,7 **(2017)** 3599-3610.

64. K. Shimizu, H. Kawachi, A. Satsuma, *Study of active sites and mechanism for soot oxidation by silver-loaded ceria catalyst,* Appl. Catal. B, 96 **(2010)** 169-175.

65. G. Corro, U. Pal, E. Ayala, E. Vidal, *Diesel soot oxidation over silver-loaded SiO2 catalysts,* Catal. Today, 212 **(2013)** 63-69.

66. K.Yamazaki, T. Kayama, F. Dong, H. Shinjoh, *A mechanistic study on soot oxidation over CeO2-Ag catalyst with 'rice-ball' morphology,* J. Catal. 282 **(2011)** 289-298.

67. S. Imamura, H.i Yamada, K. Utani, *Combustion activity of Ag/CeO2 composite catalyst*, Appl. Catal. A: Gen. 192 **(2000)** 221-226.

68. Z. Qu, F. Yu, X. Zhang, Y. Wang, J. Gao, *Support effects on the structure and catalytic activity of mesoporous Ag/CeO2 catalysts for CO oxidation*, Chem.Eng. J 229 **(2013)** 522-532.

69. X. Zhang, Z. Qu, F.i Yu, Y. Wang, X. Zhang, *Effects of pretreatment atmosphere and silver loading on the structure and catalytic activity of Ag/SBA-15 catalysts,* J. Mol. Catal. A 370 **(2013)** 160-166.
70. Q. Ye, J. Zhao, F. Huo, J. Wang, S. Cheng, T. Kang, H. Dai, Q.Ye, J. Zhao, F.Huo, J. Wang, S. Cheng, T. Kang, H. Dai, *Nanosized Ag/a-MnO$_2$ catalysts highly active for the low-temperature oxidation of carbon monoxide and benzene*, Catal.Today 175 **(2011)** 603- 609.
71. M. Skaf, PhD thesis, *Comparaison physicochimique et des activités catalytiques dans les réactions d'oxydation entre deux series de catalyseurs Ag/CeO2 préparés par impégnation et dépôt-précipitation* **(2013),** Université du Littoral cote d'opale and University of Balamand, France , Dubai.
72. T. Barakat, PhD thesis, *Oxidation of VOCs in the presence of Au and/or Pd catalysts deposited on doped nanostructured TiO2* **(2012).**
73. A. Aranzabal, J.L. Ayastuy-Arizti, J.A. González-Marcos, and J.R. González-Velasco, *The reaction pathway and kinetic mechanism of the catalytic oxidation of gaseous lean TCE on Pd/alumina catalysts*, J. Catal. 214 **(2003)** 130-135.
74. G. Veser, M. Ziauddin , L. D. Schmidt, *Ignition in alkane oxidation on noble-metal catalysts*, Catal. Today 47 **(1999)** 219-228.
75. P. Mars, D.W.Van Krevelen, *Oxidations carried out by means of vanadium oxide catalysts*, Chem. Eng. Sci 3 **(1954)** 41-59.
76. C. Doornkamp, V. Ponec, *The universal character of the Mars and Van Krevelen mechanism*, J. Mol. Catal. A: Chem, 162 **(2000)** 19-32.
77. A. Biabani-Ravandi, M. Rezaei, Z. Fattah, *Catalytic performance of Ag/Fe2O3 for the low temperature oxidation of carbon monoxide,* Chem. Eng. J. 219 **(2013)** 124-130.
78. M. Popova, A. Szegedi, Z. Cherkezova-Zheleva, A. Dimitrova, I. Mitov, *Toluene oxidation on chromium- and copper-modified SiO2 and SBA-15*, Appl. Catal. A: Gen. 381 **(2010)** 26-35.
79. P. Papaefthimiou, T. Ioannides, X. E. Verykios, *VOC removal: investigation of ethylacetate oxidation over supported Pt catalysts.* Catal. Today, 54 **(1999)** 81-92.
80. K. Okumura, T. Kobayashi. H. Tanaka, M. Niwa, *Toluene combustion over palladium supported on various metal oxide supports*, Appl. Catal. B: Environ 44 **(2003)** 325-31.
81. C. Lamonier, J.F. Lamonier, B. Aellach, A. Ezzamarty, J. Leglise, *Specific tuning of acid/base sites in apatite materials to enhance their methanol thiolation catalytic performances*, Catal. Today 164 **(2011)**124-30.
82. M. F. Finol J. Q. Torres, J. M. Giraudon, A. Gervasini, J.F. Lamonier, *Formaldehyde oxidation over hydroxyapatites: influence of the acid base properties*, Europa. Cat X, Glasgow (Scotland) **(2011).**
83. INRS, Fiche toxicologique n°74 *"Toluène"*, **2004** edition.
84. INERIS, *Données technico-économiques sur les substances chimiques en France* : toluène, 43 p. (http://rsde.ineris.fr/ or http://www.ineris.fr/substances/fr/ **(2011).**
85. INRS, Fiche toxicologique n°74 "Toluène", edition **(2012).**
86. A. Picot and F. Montandon, , *Ecotoxicochemistry applied to hydrocarbons*, chapter **(2013).**
87. H. J. Kim, S. W.Choi and H. I. I Nyang, *Catalytic oxidation of touene in contaminant emission control systems usingMn-Ce/γ-Al2O(3),* Environ. Technol 29 **(2008)** 559-569.
88. S.C. Kim, W. G. Shim, *Catalytic combustion of VOCs over a series of manganese oxide catalysts*, Appl. Catal. B: Environ. 98 **(2010)** 180-185.

89. K. Jiratova, J. Mikulova, J. Klempa, T. Grygar, Z. Bastl, F. Kovanda, *Modification of Co-Mn-Al mixed oxide with potassium and its effect on deep oxidation of VOC*, Appl. Catal.A: Gen. 361 **(2009)** 106-116.
90. M. Hosseini, T. Barakat, R. Cousin, A. Aboukais, B.-L. Su, G. De Weireld, S. Siffert, *Catalytic performance of core-shell and alloy Pd-Au nanoparticles for total oxidation of VOC: The effect of metal deposition*, Appl. Catal. B 111 **(2012)** 218-224.
91. M. Hosseini, S. Siffert, H.L. Tidahy, R. Cousin, J.-F.Lamonier, A. Aboukais, A. Vantomme, M. Roussel, B.-L. Su, *Promotional effect of gold added to palladium supported on a new mesoporous TiO2 for total oxidation of volatile organic compounds,* Catal. Today 122 **(2007)** 391-396.
92. J. G. Guo, Z. Li, H. X. Xi, C. He, B. G. Wang, J. Chem. Eng. of Chinese Universities, 19 **(2005)** 776-780.
93. X. Liang , F. Qi , P. Liu, G. Wei , X. Su, L. Ma, H. He, X. Lin, Y. Xi, J. Zhu, R. Zhu, Performance of Ti-pillared montmorillonite supported Fe catalysts for toluene oxidation: The effect of Fe on catalytic activity, Appl .Clay. Sci. **(2016).** http://dx.doi.org/10.1016/j.clay.2016.05.022.
94. M. Xue, J. Ge, H. Zhang, J. Shen**,** *Surface acidic and redox properties of V-Ag-Ni-O catalysts for the selective oxidation of toluene to benzaldehyde*, Appl. Catal.A: Gen *330* **(2007)** 117-126.
95. H. Ge, G. Chen, Q. Yuan, H. Li, *Gas phase partial oxidation of toluene over modified V2O5/TiO2 catalysts in a microreactor,* Chem. Eng. J. 127 **(2007)** 39-46.
96. Z. Qu, Y. Bu, Y. Qin, Y. Wang, Q. Fu, *The improved reactivity of manganese catalysts by Ag in catalytic oxidation of toluene*, Appl. Catal. B: Environmental 132133 **(2013)** 353-362.
97. J. Li, Z. Qu, Y. Qin, H. Wang, *Effect of MnO2 morphology on the catalytic oxidation of toluene over Ag/MnO2 catalysts*, **(2016).** http://dx.doi.org/doi:10.1016/j.apsusc.2016.05.114.
98. X .D. Zhang , L. Song, F. Bi, D.F. Zhang, Y. X. Wang, L.F. Cui, *Catalytic oxidation of toluene using a facile synthesized Ag nanoparticle supported on UiO-66 derivative*, J Colloid Interf Sci. 571 **(2020)** 38-47.
99. M. Lu, W. Yang, C. Yu, Q. Liu, D. Ye, *Plasma-catalytic Oxidation of Toluene on Ag ModifiedFeOx/SBA-15,* Aerosol Air Qual Res 20 **(2020)** 193- 202.
100.Y. Qin, Z. Qu, C. Dong, N. Huang, *Effect of pretreatment conditions on catalytic activity of Ag/SBA-15 catalyst for toluene oxidation*. Chinese J Catal, 38 **(2017)** 1603-1612.
101.J. Bedia, J.M. Rosas, J. Rodríguez-Mirasol , T. Cordero, *Pd supported on mesoporous activated carbons with high oxidation resistance as catalysts for toluene oxidation*, Appl.Catal. B: Environ 94 **(2010)** 8-18.
102.S.Velu, N.Shah, T.M. Jyothi, S.Sivasanker, *Effect of manganese substitution on the physicochemical properties and catalytic toluene oxidation activities of Mg-Al layered double hydroxides*, Micropor.Mesopor. Mat 33 **(1999)** 61-75.
103.S. Scirè ,S. Minicò, C. Crisafulli, C. Satriano, A. Pistone, *Catalytic combustion of volatile organic compounds on gold/cerium oxide catalysts*, Appl. Catal. B: Environ 40 **(2003)** 43-49.
104.J. Deng, S. He, S. Xie, H. Yang, Y. Liu, G. Guo, H. Dai, *Ultralow Loading of Silver Nanoparticles on Mn2O3 Nanowires Derived with Molten Salts: A HighEfficiency Catalyst for the Oxidative Removal of Toluene*, Environ. Sci. Technol. 49 **(2015)** 11089-11095.
105.S.C. Kim, and J. Y. Ryu, *Properties and performance of silver-based catalysts on the*

catalytic oxidation of toluene, Environ.Technol. 32 **(2011)** 561-568.
106.D.Yu, Y. Liu , Z. Wu, *Low-temperature catalytic oxidation of toluene over mesoporous MnOx-CeO2/TiO2prepared by sol-gel method,* Catal.Commun. 11 **(2010)** 788-791.
107.Z. Abbasi , M. Haghighi , E. Fatehifar , N. Rahemi, *Comparative synthesis and physicochemical characterization of CeO2 nanopowder via redox reaction, precipitation and sol-gel methods used for total oxidation of toluene,* Asia-Pac. J. Chem. Eng. 7 **(2012)** 868-876.
108.S. Ivanova, PhD thesis, *Formation of supported gold nanoparticles: from preparation to catalytic reactivity,* **(2004)** Université Louis Pasteur de Strasbourg, France.
109.B. Tapin, PhD thesis, **(2012)** University of Poitiers, France.
110.C. VIAZZI, PhD thesis, *Elaboration by the sol-gel process of Yttriated zirconia coatings on metal substrates for thermal barrier applications,* **(2007),** Université Toulouse III-Paul Sabatier UFR PCA.
111.J. Livage, M. Henry and C. Sanchez, *Sol-gel chemistry of transition metaloxides*, Prog . Solid. Chem, 18 **(1988)** 259-341.
112.C. Sanchez, J. Livage, M. Henry and F. Babonneau, *Chemical modification of alkoxide precursors,* J. Non-Cryst .Solids, 100 **(1988)** 65-76.
113.D. C . Bradley, R .C. Mehrotra, D. P. Gaur, *Metal Alkoxides*, Angew. Chem. 92 **(1978)** 975-975.
114.R.C.Mehrotra, *Synthesis and reactions of metal alkoxides*, J. Non-cryst.Sollids 100 **(1988)** 1-15.
115.J. C. Debsikdar,*Thermal evolution of alkoxy-dertved "glass-like" transparent zirconia gel,* J.Non-Cryst. Solids 87 **(1986)** 343-349.
116.R. Nass and H. Schmidt, *Synthesis of an alumina coating from chelated aluminiumalkoxides*, I. Non-Cryst. Solids 121 **(1990)** 329-333.
117.F. Babonneau, L. Coury, J. Livage, *Aluminuim sec-butoxide modified with ethylacetocetate,: An attractive precursor for the sol-gel synthesis of ceramics*, J. NonCrystal. Solids, 121 **(1990)** 153-157.
118.H. Unama and T. Tokoka, *Preparation of transparent amorphous tungsten trioxide thin films by a dip-coating method,* J. Mater. Sci. Lett. 5 **(1986)** 1248-1250.
119.J. D. Mackenzie, *Applications of the sol-gel process*, J. Non-Cryst. Solids, 100 **(1988)** 162-168.
120.E. Emeli, L. Incoccia and S. Mobilio, *Simple Alkoxide Based precursor system*, J. Non-Cryst. Solids 74 **(1985)** 3-28.
121.R. Caruso, Oscar de Sanctis, A. Macias-Garcia, E. Benavidez, S. R. Mintzer, *Influence of pH value and solvent utilized in the sol-gel synthesis on properties of derived ZrO2 powders,* Mater. Process. Technol, 152 **(2004)** 299-303.

Chapter II

Experimental techniques and raw materials

II.1 Introduction

The characterisation techniques (physisorption of nitrogen at 77 K, X-ray diffraction (XRD), temperature-programmed desorption by oxygen (DTP-O2) and by ammonia (DTP-NH3), XPS spectroscopy, temperature-programmed reduction by hydrogen (RTP-H2) and UV-Visible spectroscopy, the set-up for the toluene (C7H8) oxidation reaction, the products and the equipment used are presented in this chapter.

II.2 Materials and products used

The equipment and products used are shown in Tables 9, 10 and 11.

II.2.1 Chemical products

Table 9. Products used for sample preparation.

Chemical product	*Origin*	*Purity (%)*
Zirconium butoxideZr(OC4H_9)$_4$	Aldrich	80
Ethanol absolute(C2HO)	Aldrich	99,80
Ethanol 96%(C2H_6O)	Aldrich	96,00
Nitric acid(HNO3)	CDMChemicals Developping&Manufacturing	65,00
Ethylacetoacetate (C6H10O3)	Fluka	> 99,00
Silver nitrate (AgNO3)	Aldrich	> 99,99
Cerium (III) nitrate hexahydrate (CeN3O_9 6H2O)	Aldrich	-
Iron (III) nitrate ($Fe(NO_3)_3$, $6H_2O$)	Aldrich	>99,95

Table 10. Product used in the catalytic test.

Chemical product	*Origin*	*Purity (%)*
Toluene (C7H8)	Aldrich	anhydrous, 99.8

Table 11. Gases used in the catalytic test.

Type of gas	*Origin*	*Purity (%)*	*Use*
O2	Air Liquide	99	Calcination
20% O2	Air Liquide	>99,99	Oxidation C7H_8
H2/Ar (5/95, Vol/Vol)	Air Liquide	99,99	RTP-H2
He	Air Liquide	99,99	DTP+ test
O2/H (10/90, Vol/Vol)	Air Liquide	99,95	DTP-O2 +Test
NH3/H (5/95, Vol/Vol)	Air Liquide	99,95	DTP-NH3

II.2.2 Drying and calcination of prepared solids

The ZrO2, yCe-ZrO2 and y Fe-ZrO2 catalyst supports are dried under supercritical solvent conditions using a Prolabo autoclave with stainless steel walls. It has a volume of 1 L and can operate at up to 350°C under a pressure of 340 bar (Figure 2). Drying is carried out as follows: two Pyrex glass beakers, the first containing the gel is immersed in the second containing the ethanol and placed in the autoclave, sealed with a silver metal gasket secured by six screws, and heated in a thermostatically-controlled tube furnace. The supercritical temperature of the solvent is set by the temperature programmer. Heating causes an increase in pressure. The system is kept invariant for 15 min under hypercritical conditions, then the solvent is removed by a needle valve to lower the atmospheric pressure again.

Figure 2. Prolabo autoclave.

The samples were placed in a quartz reactor and calcined at 550°C for three hours under a stream of oxygen (30 $cm^3.min^{-1}$), the temperature rise being 1°C. min⁻. The calcination set-up is shown in Figure 3 B.

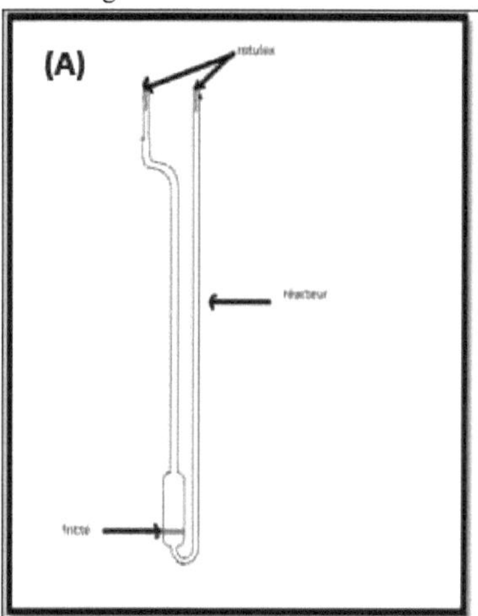

Figure 3: (A) Quartz reactor and (B) Calcination set-up.

II.3 Sample characterisation techniques

II.3.1. Physisorption of nitrogen at -196°C

11. 3. 1.1. Principle of the method

This technique is used to study the specific surface area and pore size of solids. It involves determining an isotherm that relates, at the nitrogen liquefaction temperature (-196°C), the quantity of nitrogen adsorbed at equilibrium on the surface of a solid that has already been degassed, to the pressure P of nitrogen in the gas phase. The quantity of N2 adsorbed is plotted as a function of the relative pressure P/P° (P° is the saturation pressure of nitrogen at -196°C).

The volume of nitrogen measured when P/P° is increased or decreased gives the adsorption and desorption isotherms, which are of three types according to the IUPAC classification [1] (Figure 4).

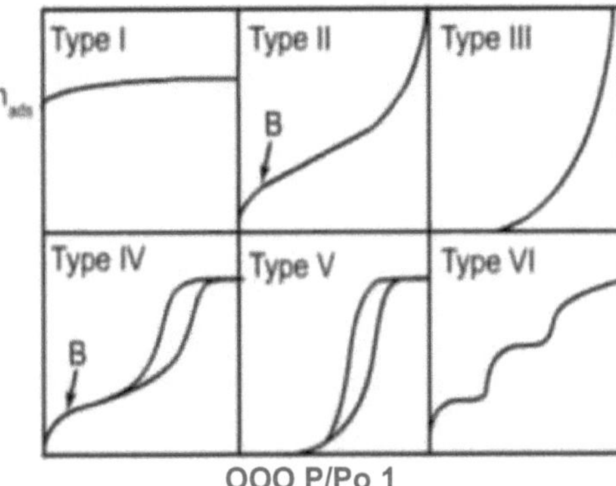

Figure 4. The different types of N2 adsorption-desorption isotherms [2].

- The type (I) isotherm corresponds to microporous (pore diameter less than 20 Â) or non-porous solids.
- Type (II) and (III) isotherms correspond to macroporous solids (pore diameter greater than 500 Á). Isotherm III is rarely observed and corresponds to a very low adsorption enthalpy.
- Type (IV) and (V) isotherms characterise mesoporous solids (diameter between 20 and 500 Á). Like isotherm III, isotherm V is rarely found.
- *Walking isotherms (type VI)* are rare and are obtained for compounds with well-defined surfaces, such as graphite carbon black.

The type of hysteresis that appears on the N2 adsorption-desorption isotherms gives information about the shape of the pores. The IUPAC hysteresis types are shown in Figure 5.

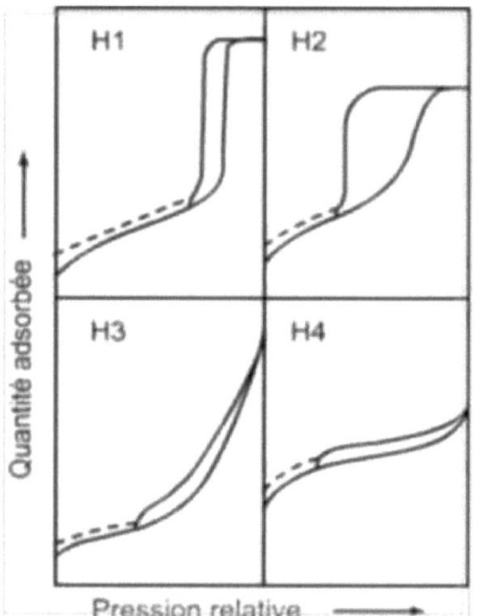

Figure 5. The different types of hysteresis [3].

- *The H1 type hysteresis loop* is often obtained and describes cylindrical pores of constant cross-section.
- *The H2 type hysteresis loop* corresponds to intergranular porosity obtained when the pore distribution is not well defined.
- *The H3 type hysteresis loop* is obtained for particle aggregates in the form of sheets and pores in the form of slits.
- *The H4 type hysteresis loop* corresponds to solids with pores in the form of narrow slots.

11. 3. 1.2. Determination of specific surface

The various methods using physical adsorption of N2 at low temperature (close to its liquefaction temperature -196°C) are based on the theory of Brunauer, Emmet and Teller (BET) [3], which is based on multimolecular adsorption. The BET model equation is given by the following relationship:

$$\frac{P}{V(P0-P)} = \frac{1}{Vm \times C} + \frac{C-1}{Vm \times C} \times \frac{P}{P0}$$

- P: Equilibrium pressure.
- P0: Saturation vapour pressure.
- V: Volume of gas adsorbed at equilibrium pressure.
- Vm: Volume of adsorbed gas corresponding to a monolayer.
- P/P0: Relative pressure.
- C: BET constant related to the heat of adsorption.

$$C = \exp{(Ea - El)} / RT$$

Ea: Heat of adsorption of the first layer ($kJ.mol^{-1}$).
El: Heat of liquefaction ($kJ.mol^{-1}$).
R: Perfect gas constant ($kJ.mol^{-1}$).
T: Temperature (K).

This BET equation applies for values of P/P0 between 0.05 and 0.35. By plotting the values of P/V (P0 - P) as a function of P/P0, we obtain a straight line whose slope and y-intercept allow us to determine the volume of the monolayer Vm, which corresponds to the quantity of gas required to cover the surface of a monomolecular layer.

$$\mathbf{Vm = 1/(a+b)}$$

Where: a: slope of the straight line P/V (P0 -P) = f (P/P0) and b: y-intercept.
The specific surface area (SBET) is obtained by this formula:

$$S_{BET}\,(m^2.g^{-1}) = \frac{\sigma.Vm.N}{V0}$$

With:
Vm= Volume of adsorbed gas corresponding to a monolayer.
N= Avogadro number (6.023 10^{23} mol^{-1}).
σ= Surface area occupied by a gas molecule i.e. $16.2.10^{-20}$ m^2 for a nitrogen molecule
V0= Volume occupied by one mole of gas (22414 $cm^3.mol^{-1}$)
So :

$$\mathbf{S_{BET} = 4{,}35\ Vm}$$

II. 3. 1.3. Determination of pore distribution

The Barrett, Joyner and Halenda [4] (B.J.H) method is based on a step-by-step analysis of the N2 desorption branch, starting from the highest relative pressure reached. Intervals of relative pressure are defined and it is assumed that, at each point, the desorbed gas comes, on the one hand, from the desorption of the gas in a certain range of pore sizes rk (the smaller the range, the lower the pressure) and, on the other hand, from the reduction in thickness (dt) of the film adsorbed on the walls of the larger pores previously emptied of their condensed gas.
The B. J. H. method is based on Kelvin's empirical law as follows :

$$\mathbf{Ln\,(P/P0) = -\,2.\,\gamma\,.Vm\,/\,R.T.rk}$$

With :
γ: Surface tension of the condensed liquid.
Vm: Molar volume of the condensed liquid.
rk: Pore radius (Á).
The pore radii are determined by the following formula:

$$\mathbf{r_k = -K\,/\,Ln\,(P/P_0)}$$

With :
K is a constant equal to 4.10 when the adsorbate is nitrogen.
On this branch of the curve, we can determine two very close volumes V1 andV2 as well as the two corresponding relative pressures $(P/P0)_1$ and $(P/P0)_2$ and therefore two pore radii r1 and r2, which allows us to plot the curve dV/drp = f (rp).

11. 3. 1.4. Determination of porous and microporous volumes

The total pore volume of the sample corresponds to the volume of gas adsorbed at the relative pressure P/P0 = 0.95.
The microporous volume is determined using De Boer's t-plot method [5], which consists of

comparing the nitrogen adsorption isotherms obtained on porous solids with the standard isotherm determined on non-porous solids, which represents a straight line. Thus, by plotting the value of the adsorbed volume V as a function of the thickness t of the layer, determined by multiplying the thickness of a monolayer by V/Vm, we obtain an isotherm which has a linear part parallel to the standard isotherm. Extrapolation of this straight line gives the microporous volume directly.

12. 3.1.5 Equipment and sample preparation

Analysis by nitrogen physisorption at -196°C was carried out using a Micrometrics ASAP 2020 automated instrument (Figure 6). This instrument has two compartments: a left-hand compartment for degassing the sample and a right-hand compartment for carrying out the analysis. The experimental device is fitted with a system of valves allowing the two compartments to be connected.

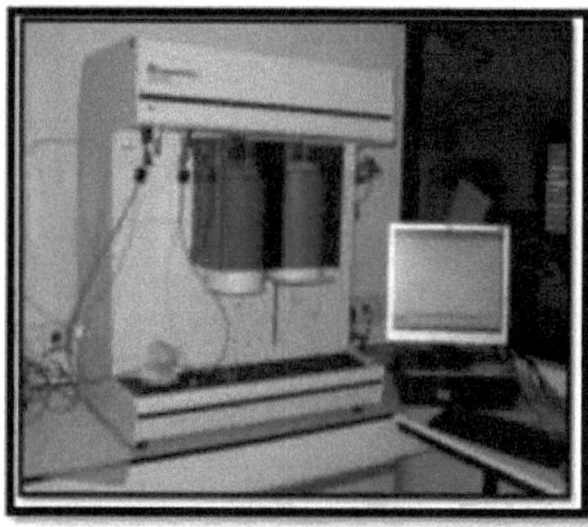

Figure 6. Diagram of the Micrometrics ASAP 2020 instrument.

Before analysis by N_2 physisorption, the sample (50 mg) was degassed at 250°C for 5 h.

II. 3.2. X-ray diffraction

II. 3.2.1 Principle of the XRD method

This method of analysis, discovered by Max Von Laue (Nobel Prize in 1914) and widely studied by Sir William Henry Bragg and his son Sir William Lawrence Bragg (joint Nobel Prize in 1915), enables the crystallinity of solids to be studied.

XRD involves applying X-ray wavelength radiation (0.1 nm $< \lambda <$ 10 nm) to a sample. The radiation penetrates the solid, absorbing some of the energy and exciting the atoms, emitting radiation in all directions. Radiation emitted by atomic planes that are in phase generates a coherent beam that can be detected [6] .

Bragg's law is as follows:

$$\mathbf{2\,d\,\sin\Theta = n\,\lambda}$$

With:

d = Inter-reticular distance (in Å).

θ = Angle of diffraction (degrees).

n = diffraction order.

λ = X-ray wavelength.

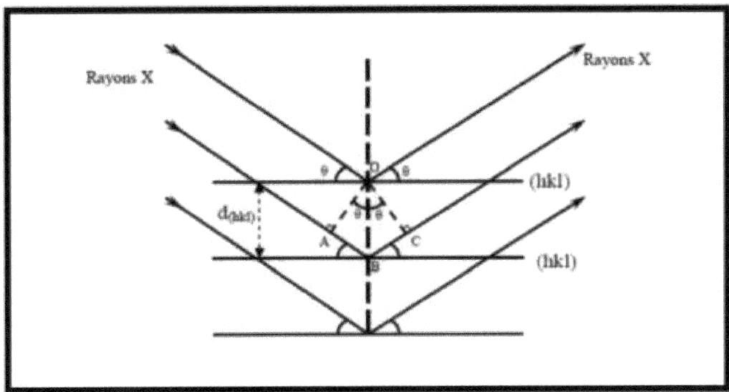

Figure 7. Diagram explaining Bragg's law.

11. 3. 2.2. Equipment and sample preparation

The XRD analyses were carried out at room temperature on a BRÜKER D8 Advance diffractometer (Figure 8). This diffractometer is automated and equipped with a copper anti-cathode (λ= 1.54184 Â). The sample to be analysed is placed in powder form on a flat support. The general acquisition conditions correspond to an angular range in 2θ from 15 to 70° with a measurement step of 0.02° and an integration time of 2 s.

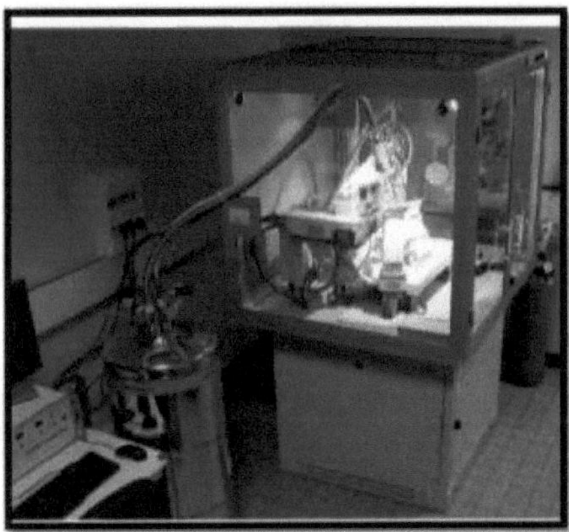

Figure 8. Brüker AXS D8.

12. 3.3. UV-visible spectroscopy

13. 3.3.1 Principle of the method

Light absorption in the ultraviolet and visible ranges causes valence electrons to jump between the various molecular orbitals, providing information about the electronic structure [7].

There are different types of electronic transitions:

- d-d transitions (for transition metal ions).
- Π-Π^* transitions (for organic molecules)
- Charge transfer (transfer of electrons to an unoccupied orbital)

The measurement of different electronic transitions in transmission in the case of liquids and in diffuse reflection in the case of solids (DRUV-Vis).

II. 3. 3.2. Analysis conditions and equipment

A Perkin Elmer model lambda 40 UV Visible spectrometer (Figure 9) was used for diffuse reflection UV-vis analysis. The spectra were recorded in reflectance mode in the wavelength range 200-800 nm. The solid to be analysed is packed into a sample holder with a spectrasil window. The reference used is $BaSO_4$.

Figure 9. PERKIN ELMER LAMBDA 40 UV-vis spectrometer.

II.3.4. X-ray photoelectron spectroscopy (XPS)

II.3.4.1. Principle of the method

X-ray photoelectron spectroscopy (XPS) is a non-destructive characterisation technique based on the photoelectric effect, one of the processes by which photons interact with atomic electrons [8].

XPS measurement is particularly suitable for characterising all the chemical elements (except hydrogen and helium) present on the surface of a material. This is because sending radiation (hv) whose energy can extract hollow electrons (and not valence electrons) from the atoms on the surface of the material. The sample to be analysed is bombarded with electromagnetic radiation of hv energy in the X-ray range (1 to 2 keV). The photons absorbed by the atoms give rise to an emission of electrons, known as photoelectrons. These photoelectrons come from the atomic energy levels of the various elements and are analysed in terms of number and energy by an appropriate detector.

X-ray photoelectron spectroscopy is based on the principle of conservation of the energy of the incident photon and is expressed as follows:

$$h\nu = E_c + E_L + \varphi_{ech}$$

Where *hv*: Energy of a photon (in J)

h: Planck's constant (in J.s) (~6.626 0 755x10-34)

v: Radiation frequency (in s^{-1})

EL: Binding energy (in J)

E_c*:* Electronic kinetic energy (in J)

феей: the sample output work.

The spectrometer output work (феей) as well as the incident photon energy (hv) are known parameters of the experimental setup.

Let *Ec*= hv - *Ei,* Ei = Ef - Ei where Ei and Ef are the total energies of the system in the initial (N electrons) and final (N-1 electrons) photoemission states.

> This technique provides access to information on :

- The chemical composition and relative concentration of the various constituents of the sample in the first surface layers (5-50 Â).
- The chemical environment of the different elements (distinction between a metal and its oxides).
- The electronic structure and oxidation state of atoms.

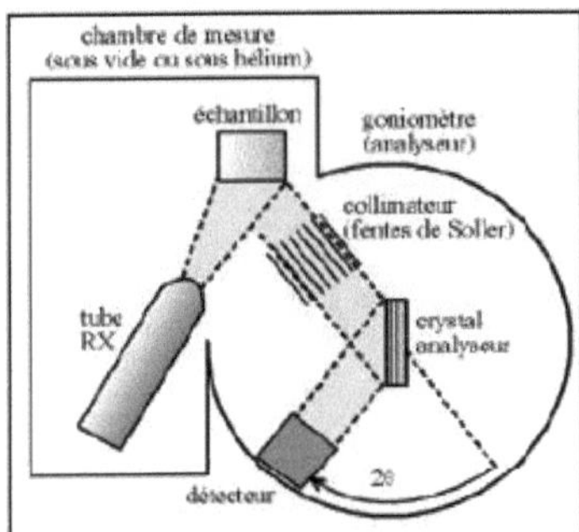

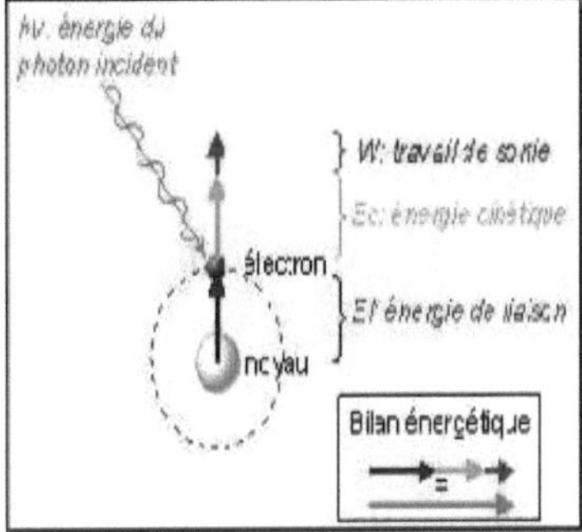

Figure 10. Principle and operation of X-ray photoelectron spectroscopy.

1.1.1.2. Equipment and sample preparation

The XPS apparatus consists of an X-ray source, an irradiation chamber, an ion gun, a charge neutraliser, pumps, an electron analyser and a detector. The sample, which is in powder form and whose surface must be kept free of contamination, is placed on a support and then introduced into a first chamber (known as the entry chamber, which is evacuated to pressures of less than 10-7 mbar using a secondary pump). Once this vacuum has been reached, the sample is transferred to the analysis chamber for XPS analysis, which lasts between one and four hours on average. The spectra obtained are processed by computer.

The X-ray photoelectron spectra were produced using an automated ESCALAB 250 spectrometer. The source consists of an aluminium anode emitting monochromatic X-rays (Ka = 1486.6 eV).

1.1.1.3. 5. Hydrogen programmed temperature reduction (RTP-H2)

1.1.1.4. . Principle of the method

This catalyst characterisation technique is based on the determination of hydrogen consumption by reduction as a function of temperature. This makes it possible to determine the reduction temperature of the species present in the materials analysed, and to obtain other information such as the stability and quantity of these species.

1.1.1.5. Equipment and sample preparation

The apparatus used to study the reducibility of materials is a Micromeritics Auto Chem 2910 (figure 11). The apparatus consists of a thermal conductivity detector (used to detect and quantify the quantity of hydrogen), a gas system consisting of mass flow meters and several three-way and six-way valves (used to precisely control the flow and circulation of the gas stream), an oven fitted with a thermocouple (to regulate the temperature rise of the sample), a saturator fitted with a heating mantle, a water trap (to protect the detector from the water formed during reduction) and a zero station (to calibrate the consumption of oxygen or hydrogen).

RTP-H2 analysis was carried out according to the procedure reported in reference [9]. The 80 mg sample was initially pre-treated in situ at 550°C under a flow of 5 vol % $_{O2/He}$ (30 cm^3/min) for 30 min, then brought back to 50°C and purged under He (30 cm^3/min) before undergoing a programmed temperature rise from 50°C to 800°C (10°C/min) under a reducing gas mixture (5% $_{H2/Ar}$) at a flow rate of 30 cm^3/min.

Figure 11. Diagram of the Auto Chem 2910.

11.3.6. Temperature-programmed desorption using ammonia (DTP-NH3)

11.3.6.1. Principle of the method

This method is based on the thermodesorption at increasing temperature of NH3 pre-adsorbed on a solid, which makes it possible to assess the total acidity and provides information on the strength and distribution of the acid sites.

11.3.6.2. Apparatus and sample preparation

The instrument used to study NH3 desorption is a Micromeritics Auto Chem 2910 (figure 11). The procedure used for DTP-NH3 analysis is similar to that reported in reference [9]. The 80 mg sample was pre-treated in air (30 mL/min) for 2 h at 550°C at a rate of 10°C/min. It was then cooled to 100°C under air flow and NH3 adsorption (5% $_{NH3/He}$) was carried out at this temperature for 30 min at a flow rate of 30 cm^3/min. After evacuation of the physisorbed ammonia, carried out under a flow of Helium (30 cm^3/min) for 30 min, NH3 desorption is carried out by heating the sample from 100°C to 550°C (10°C/min) under a flow of Helium (30 cm^3/min).

11.3.7. Temperature-programmed desorption with oxygen (DTP-O2)

11.3.7.1. Principle of the method

This method is based on the thermodesorption, at increasing temperature, of O2 pre-adsorbed on a solid, which makes it possible to study the interactions between the metal and the oxygen and provides information on the nature of the adsorbed oxygen species.

11.3.7.2. Apparatus and sample preparation

The instrument used to study O2 desorption is a Micromeritics Auto Chem 2910 (figure 11). The sample (mass 80 mg) undergoes the same pre-treatment as described for DTP-NH3. It was then cooled to 80°C in air (5% O2/He) at a flow rate of 15 cm^3/min and purged in helium (15 cm^3/min) for 17 min. Finally, the O2 was desorbed by heating the sample from 80°C to 550°C (10°C/min).

II.4. Catalytic oxidation test for toluene (C6H5-CH3)

11.4.1. Apparatus and sample preparation

The catalytic test is carried out under atmospheric pressure in a quartz reactor containing the sample and connected to a glass saturator containing the toluene. The test set-up is made entirely of stainless steel and is illustrated in Figure 12.

Before each reaction, a 20 mg mass of catalyst, previously calcined at 550°C under O2, was activated under a flow of oxygen (10% O2/He) for 1h at 200°C. The catalyst is then swept with a gas mixture consisting of toluene, O2, He (with [toluene]= 1000 ppm) and H2O (3.4%). Helium is used as the balance gas. Toluene oxidation was studied over a temperature range from 200°C to 550°C. The temperature was programmed to rise at a rate of 6°C/min using a temperature controller that controlled the temperature at the catalyst by means of a thermocouple placed inside the reactor.

Figure 12. Set-up for toluene oxidation.

11.4.2. Identification of reagents and products: mass spectrometry

At the reactor outlet, the gaseous effluents were analysed using a Pfeiffer Omnistar mass spectrometer (MS). The evolution of the ionic current intensity for different m/e values, corresponding to the ionic fragments of the reactants and products: toluene (92), CO2 (44), H2O (18), benzene (78), butane (44) and CO (28), was recorded during the catalytic reaction using QUADSTAR software.

11.4.3. Using the results

Total conversion of (C_7H_8) (%) $= \frac{[Toluène]0 - [Toluene]T}{[Toluène]0} \times 100$

With

[Toluene] $_0$: Initial toluene concentration (at reactor inlet).

[Toluene]$_T$: Concentration of toluene at a temperature T (at the reactor outlet).

CO2] concentration (ppm $= \frac{[I]t - [I]0}{[I]7000} * 7000$

With

[1] t: CO_2 intensity at temperature t

[2] 0: CO_2 intensity at reactor inlet

[3] 7000: CO_2 intensity at 7000

7000: initial CO_2 concentration

It is important to note that CO_2 is the only product detected on some catalysts. For others, it is mainly obtained in the presence of traces of benzene and a compound from the C3-C6 paraffin family.

11. 5. References

1. K. S. W. Sing, R.T. Williams, *Physisorption Hysteresis Loops and the Characterization of Nanoporous Materials*, Adsorption sciences-Technonolgy **(2004)** https://doi.org/10.1260/0263617053499032

2. C. Kosanovic, A. Cizmek, B. Subotic, I. Smit, *Mechanochemistry of zeolites: Part 2. Change in particulate properties of zeolites during ball milling*, Zeolites. 15 **(1995)** 247-252.

3. S.Brunauer, P. H. Emmet, E. Teller, *Adsorption of Gases in Multimolecular LayersAm.* Chem. *Soc.* 60 **(1938)** 309-319.

4. E. P. Barrett, L. G. Joyner and P. P. Halenda, T *he Determination of Pore Volume and Area Distributions in Porous Substances. I. Computations from Nitrogen Isotherms*, J. Am. Chem. *Soc.* 73 **(1951)** 373-380.

5. J. H. Deboer, B. C. Lippens, *Studies on pore systems in catalysts: V. The t-method*, J. Catal. 4 **(1965)** 319-323.

6. J. Arfaoui, *Synthèse et caractérisation de Montmorillonites pontées au Titane et Dopées au Vanadium : Application catalytique à la réduction sélective de NO par NH3 et à l'époxydation des alcools allyliques*, PhD thesis **(2010)**, Faculté des Sciences de Tunis, Tunisie

7. L. E. Orgel, Methuen, Doctoral thesis, London, **(1961)**.

8. R. Kriptal, M. Maurya, *EPR and optical absorption of VO^{2+} impurity in lithium potasium sulphate single crystal*, physica B: condensed Matter 404 **(2009)** 1532
1537.

9. J. Arfaoui, A. Ghorbel, C. Petitto, G. Delahay, *Novel V_2O_5-CeO_2-TiO_2-SO_4^{2-} nanostructured aerogel catalyst for the low temperature selective catalytic reduction of NO by NH3 in excess O_2*, Appl. Catal. B: Environ. 224 **(2018)** 264-275.

Chapter III

Synthesis and characterisation of solid materials

III.1 Introduction

This chapter presents the method for preparing the solids ZrO_2, $yCe\text{-}ZrO_2$, $yFe\text{-}ZrO_2$, Ag/ZrO_2, $Ag/yCe\text{-}ZrO_2$, $Ag/yFe\text{-}ZrO_2$, $xAg/Ce\text{-}ZrO_2$ and $xAg/Fe\text{-}ZrO_2$. The effect of varying the quantities of Fe, Ce and Ag on the physicochemical properties of these solids is studied.

III.2 Synthesis of supports and supported silver-based catalysts

III.2.1 Preparation of ZrO2 and yM-ZrO2 substrates

The method used to synthesise ZrO_2 and y $M\text{-}ZrO_2$ supports (with M= Ce or Fe) is the sol-gel route, which is based on a series of hydrolysis and condensation reactions of molecular precursors (described in Chapter I). It is used to prepare materials with advanced structural and textural properties.

The preparation of ZrO_2 is carried out using a procedure and conditions similar to those reported in references [1,2] for the synthesis of TiO_2. A specific quantity of zirconium butoxide *($Zr(OCH_2CH_2CH_2CH_3)_4$, 80%)* is added, with stirring, to a volume of pure ethanol *(C_2H_5OH)*. Next, ethylacetoacetate *($C_6H_{10}O_3$, 99%, noted Eacac)* is added as a complexing agent and the mixture is stirred at room temperature for one hour. Finally, hydrolysis was carried out by adding an aqueous solution of nitric acid *(HNO_3, 0.1 M)* and the mixture was kept under stirring until the gel was obtained.

The same procedure is used to synthesise the $M\text{-}ZrO_2$ y aerogel supports (Figure 13). To introduce the additive M, a known mass of the nitrate precursor of M (*(CeN_3O_9, $6H_2O$)* or *($Fe(NO_3)_3$, $6H_2O$)*) is added to the reaction mixture before hydrolysis. The carriers obtained are denoted $yM\text{-}ZrO_2$.

y is the mass percentage of M, where :

y = 0; 1.25; 2.5; 5; 10; 15; 20 and 30% wt. for $Ce\text{-}ZrO_2$

y = 0; 1.25; 2.5; 5 and 10% wt. for $Fe\text{-}ZrO_2$

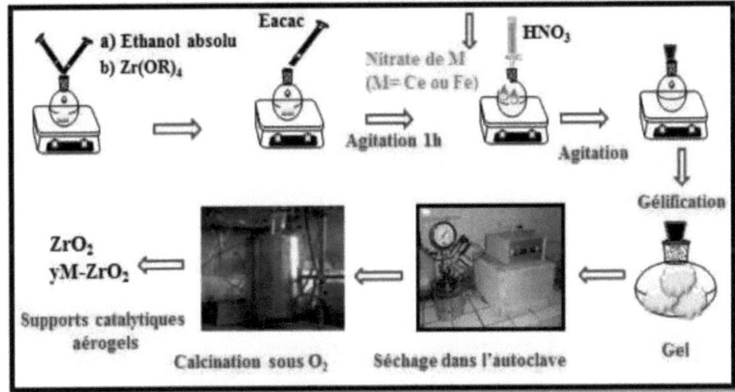

Figure 13. Different stages in the synthesis of ZrO2 and $yM\text{-}ZrO_2$ supports.

Gel drying is an important stage in the preparation of materials using the sol-gel method. An aerogel solid obtained by supercritical drying is characterised by good thermal stability, a larger specific surface area and higher porosity than a xerogel solid dried in an oven [3]. We therefore dried the gels under supercritical solvent conditions (Table 12).

Table 12. Critical temperature and pressure of ethanol [4].

Solvent	Tc (°C)	Pc (bar)
Ethanol	243	63

The solids obtained were ground and then calcined at 550°C, under a flow of oxygen (30 $cm^3.min^{-1}$) with a heating rate of 1°C. min^{-1} and a three-hour plateau.

The synthesis conditions for the solid aerogels ZrO2 and yM-ZrO2 are presented in Table 13.

Table 13. Preparation conditions for ZrO2 and yM-ZrO2 aerogels.

Catalytic supports	*ZrO2 and yM-ZrO2*
Complexation ratio	nEacac/ nZr = 1
Hydrolysis ratio (h)	nH+ / nZr = 10
Type of solid	Aerogel
Calcination temperature	550 °C

III.2.2. Synthesis of Ag/ZrO2 and Ag/yM-ZrO2 catalysts

The Ag/ZrO2 and Ag/yM-ZrO2 catalysts are synthesised by the impregnation method (Figure 14). A known mass of oxide support (ZrO2 or yM-ZrO2), prepared by sol-gel process and calcined at 550°C, is brought into contact with a small volume of an aqueous silver nitrate solution *(AgNO3, 99%)*. After stirring, the product obtained was dried overnight in an oven at 100°C and then calcined at 550°C under a flow of O2. The catalysts prepared are denoted xAg/yM-ZrO2, where x: the percentage by weight of Ag (0.5, 1, 2, 2.5 or 3% wt.).

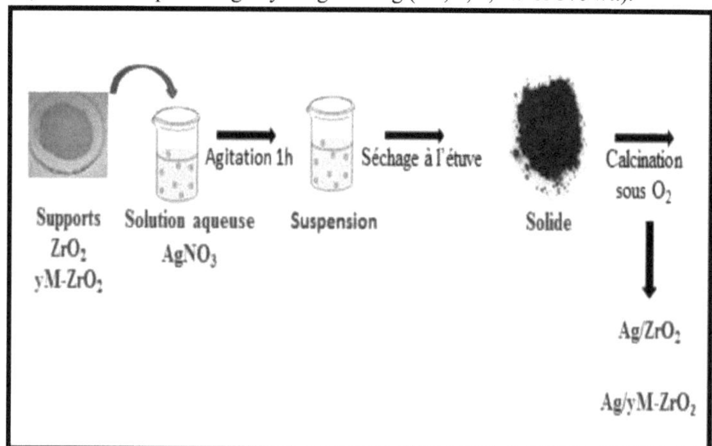

Figure 14. Different stages in the synthesis of the Ag/ZrO2 and Ag/yM-ZrO2 catalysts.

III.3 Characterisation of supports and catalysts (ZrO2, yM-ZrO2, Ag/ZrO2 and Ag/yM-ZrO2)

Determining the physicochemical characteristics of a catalyst is necessary to explain its catalytic activity. These characteristics depend on the preparation parameters; each parameter has an effect that must not be neglected. In this work, we studied the effects of the nature and quantity of the additive M and the silver content in the preparation of Ag/yM-ZrO_2 catalysts.

III.3.1 Effect of Ce content on the physicochemical characteristics of ZrO_2 and yCe-ZrO_2 solids:

III.3.1.1 XRD study

The results of the XRD structural analysis of the ZrO_2 and yCe-ZrO_2 supports (where y = 0; 1.25; 2.5; 5; 10; 15; 20 and 30% wt. Ce) are shown in Figure 15.

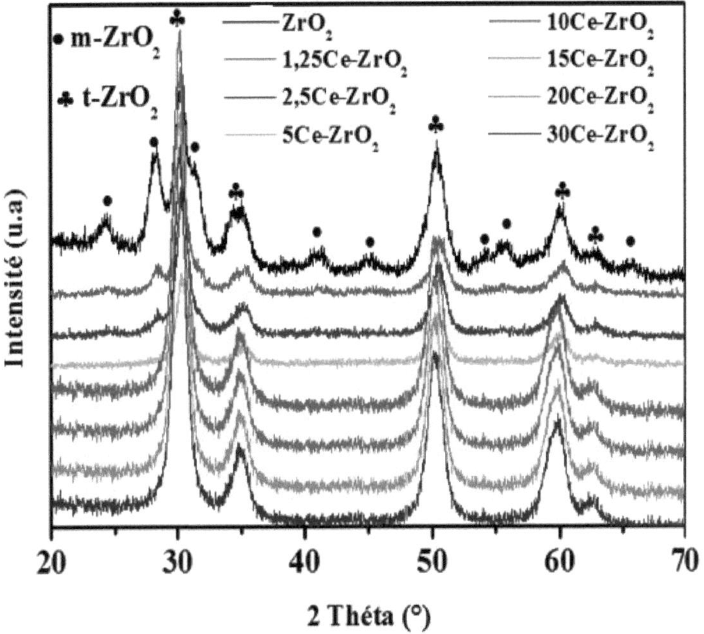

Figure 15. X-ray diffragtograms of aerogel supports (ZrO_2 and yCe-ZrO_2).

The X-ray diffractogram of the ZrO_2 support shows the lines of the two stable monoclinic (m-ZrO_2) and metastable tetragonal (t-ZrO_2) phases of zirconia. Thus the peaks located at diffraction angles 2θ = 24.38° (110), 28.26° (-111), 31.56° (111), 41.07° (102), 45.21° (-202), 54.09° (300), 55.60 (130), 65.68° (320) and 2θ = 30,33° (101), 34.67° (002), 50.37° (112), 60.09° (211), 62.57° (202) are assigned to the monoclinic [PDF 89-9066] and tetragonal [PDF 79-1769] phases of ZrO_2 , respectively.

The addition of increasing quantities of cerium modifies the structure of zirconia (Figure 15), leading to the disappearance of the characteristic lines of the monoclinic phase of ZrO_2. This shows stabilisation of the tetragonal phase of ZrO_2 (metastable at low temperature) by the addition of cerium and may suggest the existence of interactions between ZrO_2 and Ce. The stabilisation of the tetragonal phase of zirconia was also observed by W. Khaodee and V. Solinas et al. [5,6] on CeO_2-ZrO_2 catalysts.

It should be noted that the absence of CeO_2 oxide peaks in the case of all solids shows good dispersion of cerium on the ZrO_2 surface. Similar results were subsequently obtained by Z. Guerra-Que et al [7] and J. Arfaoui et al [1]. The authors showed good dispersion of Ce species on the surface of CeO_2-ZrO_2 and V_2O_5-CeO_2- TiO_2-SO_4^{2-} catalysts, respectively.

III. 3. 1. 2. Study by nitrogen physisorption at 77 K

The results of the N2 physisorption texture analysis at 77 K obtained for the yCe-ZrO2 supports are shown in Table 14.

Table 14. Results of the analysis of ZrO2 and yCe-ZrO2 aerogel supports by N2 physisorption.

Sample	*SBET (m^2.g1)*	*Vp (crf.g'l)*	*Dmoy (Å)*
ZrO2	107	0,33	88
1.25Ce-ZrO2	72	0,26	81
2,5Ce-ZrO2	82	0,21	83
5Ce-ZrO_2	112	0,29	91
10Ce-ZrO2	119	0,35	87
15Ce-ZrO2	130	0,39	110
20Ce-ZrO2	129	0,73	232
30Ce-ZrO2	117	0,47	171

Where: * SBET: specific surface area, * Vp: pore volume, *Dmoy: mean pore diameter

The results given in Table 14 show that the aerogels calcined at 550°C are characterised by a large specific surface area (72< SBET< 130 m^2g^{-1}) and a large porosity (0.21 Vp< < 0.73 cm^3g^{-1}), which makes them thermally stable.

The adsorption-desorption isotherms and pore distribution curves (Figures 16 and 17, respectively) show that the aerogels produced are characterised by :

- *Type IV isotherms* (according to the IUPAC classification) and a monomodal pore distribution relative to a homogeneous mesoporous texture.
- *H1 type hysteresis loops* (cases of : 1.25Ce-ZrO2, 2.5Ce-ZrO2, 5Ce-ZrO2, 10Ce-ZrO2, 15Ce-ZrO2, 20Ce-ZrO2 and 30Ce-ZrO2) *or H2* (case of ZrO2) corresponding to cylindrical or bottle-shaped pores.

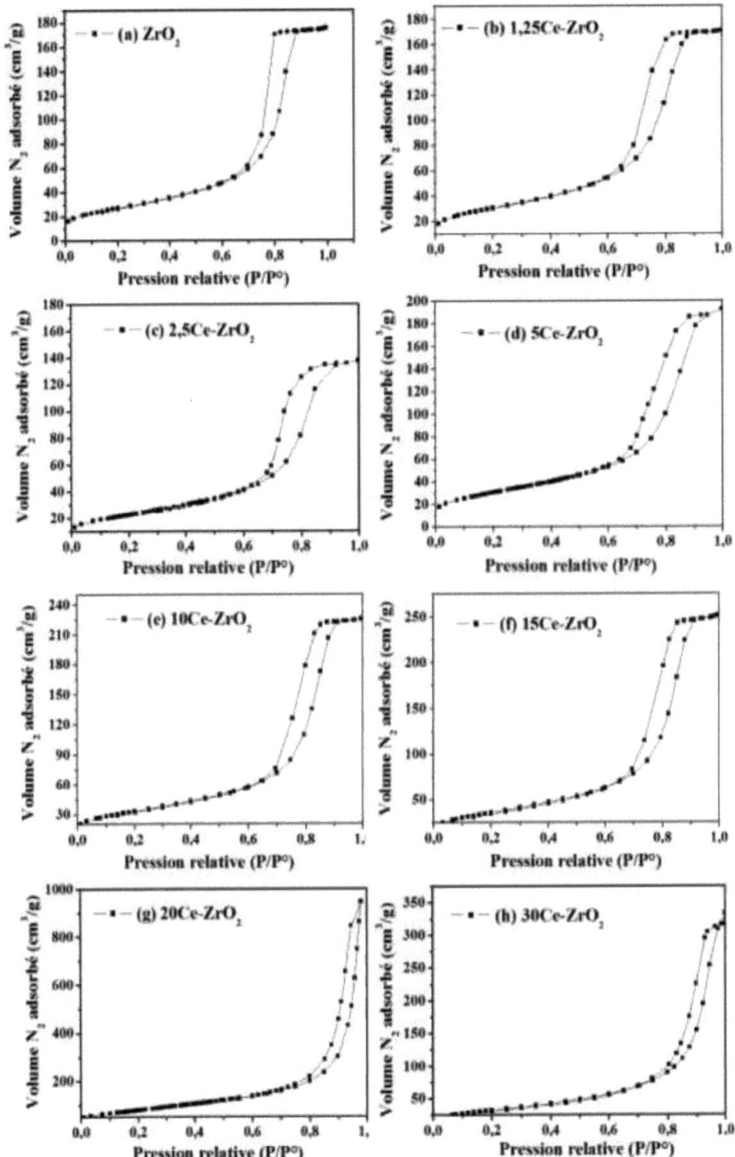

Figure 16. N2 adsorption-desorption isotherms for ZrO2 and yCe-ZrO2 supports.

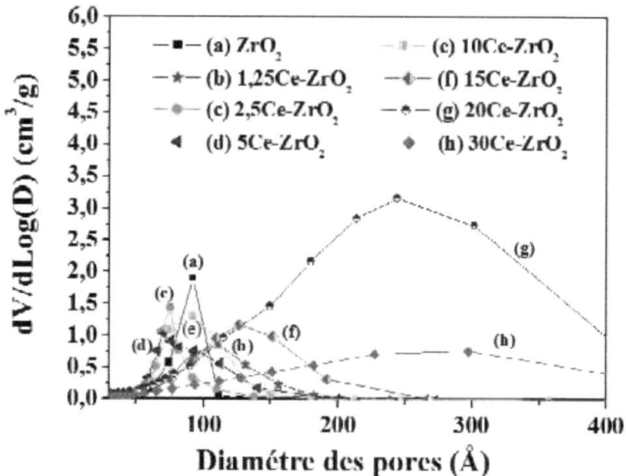

Figure 17. Pore distribution curves for ZrO2 and yCe-ZrO2 substrates.

III. 3. 1.3 UV-Visible spectroscopy

The results of the UV-Visible spectroscopy analysis of the supports (Zr and Ce) are shown in Figure 18 and listed in Table 15.

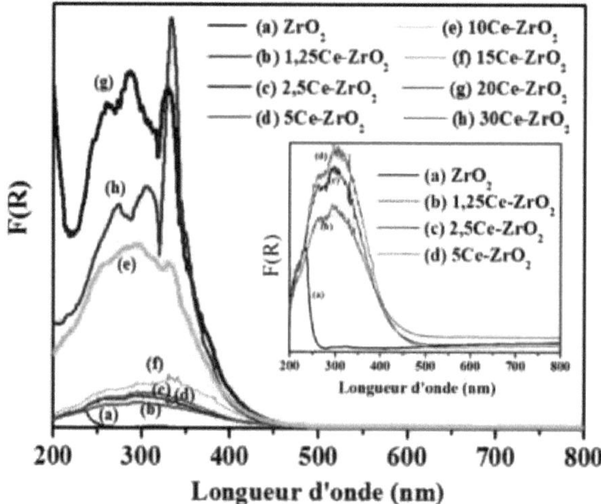

Figure 18. UV-Vis spectra of ZrO2 and yCe-ZrO2 aerogel supports.

The UV spectrum of the ZrO2 aerogel substrate (Figure 18) shows an absorption band between 200 and 250 nm attributed to the (O2- > Zr^{4+}) transition in zirconia [7,8]. For the yCe-ZrO2 samples, the new bands observed in the regions: 230-260 nm and 280350 nm relate, respectively, to the O2- > $Ce^{3+}$and O2- > Ce^{4+} transitions [1,7,9,10]. Thus, cerium species are present in the +3 and +4 oxidation states on the ZrO2 surface. This result is in agreement with those obtained previously in our research team by J. Arfaoui and C. Gannoun et al [1, 11], who detected Ce^{3+} and Ce^{4+} on the surface of the catalysts V2O5-CeO2-TiO2-SO_4^{2-} (aerogel prepared by the sol gel insitu method) and V2O5/CeO2- TiO2, (prepared by sol gel method followed by impregnation).

Table 15. UV-vis bands observed for ZrO2 and yCe-ZrO2 aerogels.

Samples	*Wavelengths (nm)*	*The corresponding bands*	*Ref. Biblio.*
ZrO2 and yCe-ZrO2	200-250	Charge transfer O2-> Zr^{4+}	[7,8]
yCe-ZrO2	230-260	Charge transferO^{2}-> Ce^{3+}	[1,7,9,10]
	280-350	Charge transferO2-> Ce^{4+}	

III. 3. 1. 4 X-ray photoelectron spectroscopy (XPS)

The XPS spectra corresponding to the Zr3d, O1s and Ce3d orbitals are shown in Figures 19a, 19b, 19c, 19d and 19e. The binding energies of these orbitals and their assignments are given in Table 16.

Table 16. Binding energies and oxidation states of atoms in the samples.

Solid	*orbitals*	*Bond energies (Ev) and oxidation state*	*Ref. Biblio.*
ZrO2 and yCe-ZrO_2	Zr 3d (3d5/2 3d3/2)	181,7 Zr^{4+} 184,5	[7,12,13]
yCe-ZrO_2	O1s	529.0-529.9 eV OL (O^{2}-) network 530-531, 9 eV O- or OH- surface chemisorbed	[14,15]
	Ce 3d ($3d_{3/2}$ and 3d5/2)	885.5 and 904.5 ^ $3d^{10}4f1$ Ce^{3+}	[1,15,16]
		900.9; 906.4; 916.9; 917.8; 882.6 and 898.3^ $3d^{10}4f0$ Ce^{4+}	[1,7,15]

The results obtained by X-ray photoelectron spectroscopy reveal the presence of $Ce^{(3+)}$) and $Ce^{(4+)}$) species, which is consistent with the results of UV-visible spectroscopy. X. Deng et al [17] also showed the presence of cerium in the +3 and +4 oxidation states on an Ag/Ce-Zr catalyst prepared by the co-precipitation method.

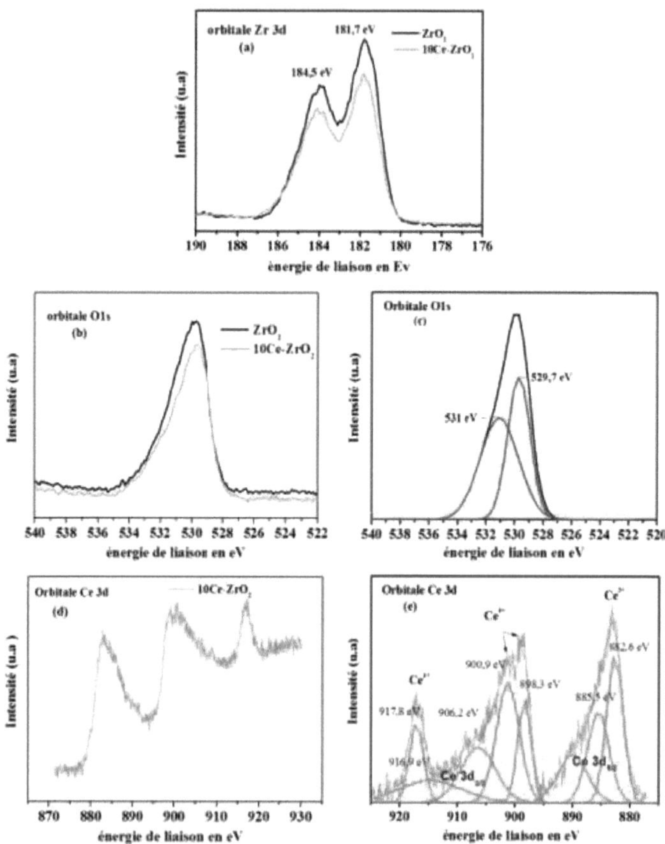

Figure 19. Results of XPS analysis of ZrO2 and Ce-ZrO2 supports.

III. 3. 1. 5 Analysis by RTP-H2

The effect of the amount of Ce on the redox characteristics of the yCe-ZrO2 solids (y = 0; 1.25; 2.5; 5; 10; 15; 20 and 30 % wt.) was studied by programmed temperature reduction using hydrogen. The results obtained are shown in Figure 20.

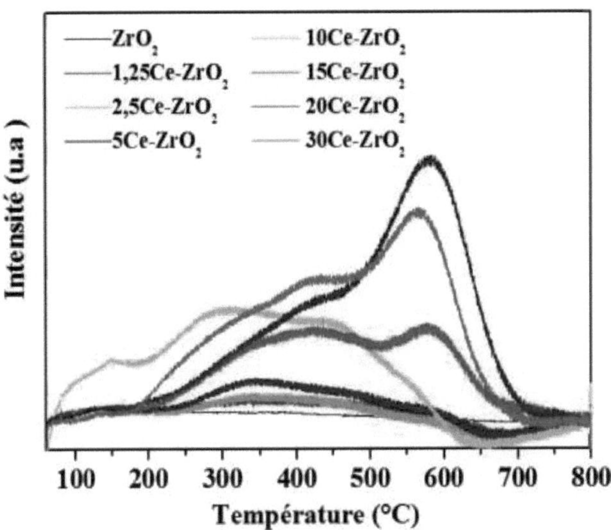

Figure 20: Results of RTP-H2 analysis of ZrO2 and yCe-ZrO2 aerogel supports.

This figure shows that zirconia does not reduce in the temperature range studied [18]. The yCe-ZrO2 solids show several H2 reduction peaks between 100 and 700°C, which are attributed to the reduction of different cerium species. According to the literature, the reduction of small species with strong interactions with the support occurs between 100 and 500°C [19-23], whereas the reduction of large species with weak interactions with the support occurs at higher temperatures (> 500°C) [19-23]. Thus, the amount of Ce probably influences the nature of the cerium species and their interactions with ZrO2.

III.3.2 Characterisation of Ag/ZrO2 and Ag/yCe-ZrO2 catalysts

III.3.2.1 XRD analysis

The results of the XRD analysis of the Ag/ZrO2 and Ag/yCe-ZrO2 catalysts (with the percentage of Ag = 2% wt) are shown in Figure 21.

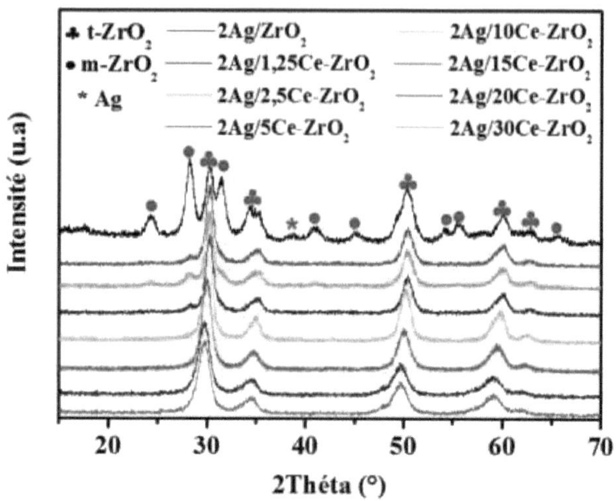

Figure 21. X-ray diffragtograms of Ag/ZrO2 and Ag/yCe-ZrO2 catalysts.

The XRD diffractograms of the Ag/yCe-ZrO2 catalysts are similar to those obtained for the yCe-ZrO2 supports (Fig. 15) hence the addition of Ag does not influence the structure of the yCe-ZrO2 aerogels. Note that a low-intensity peak relating to metallic silver appears at 2Θ = 38, 1° (111) [PDF 65-2871] only for the Ag/ZrO_2 sample. This may be related to the existence of interactions between cerium and silver that allow good dispersion of silver species. S. Imamura and Z. Qu et al [24,25] also obtained good dispersion of silver on Ag/CeO2 type catalysts.

III.3.2.2 Study by nitrogen physisorption at 77 K

The results of the analysis of the solids Ag/ZrO2 and Ag/yCe-ZrO2 by physisorption of N2 at 77 K are reported in Table 17.

Table 17. Results of analysis of Ag/ZrO2 and Ag/yCe-ZrO2 solids by physisorption of N2 at 77 K.

Samples	*SBET (trf.g'ˡ)*	*Vp (cmf.-g'ˡ)*	*Dmoy (Å)*
2Ag/ZrO₂	76	0,24	97
2Ag/1,25Ce-ZrO₂	62	0,24	148
2Ag/2.5Ce-ZrO2	77	0,22	106
2Ag/5Ce-ZrO2	81	0,25	106
2Ag/10Ce-ZrO2	73	0,31	127
2Ag/15Ce-ZrO2	118	0,38	112
2Ag/20Ce-ZrO2	68	0,33	173
2Ag/30Ce-ZrO2	101	0,47	185

Comparison of the results in this table with those in Table 14 shows that the addition of Ag leads to a slight decrease in the specific surface area of the supports, which can be attributed to the blocking of certain pores by the silver. This result is in line with numerous studies reported in the literature, which have also demonstrated a reduction in the specific surface area of different supports (Al2O3, ZrO2, TiO2, CeO2, intercalated clay, etc.) after the addition of different atoms (V, Mo, etc.) [1,25-29].

The adsorption-desorption isotherms and pore distribution curves for the Ag/ZrO2 and Ag/yCe-ZrO2 catalysts are shown in Figs. 22 and 23 respectively.

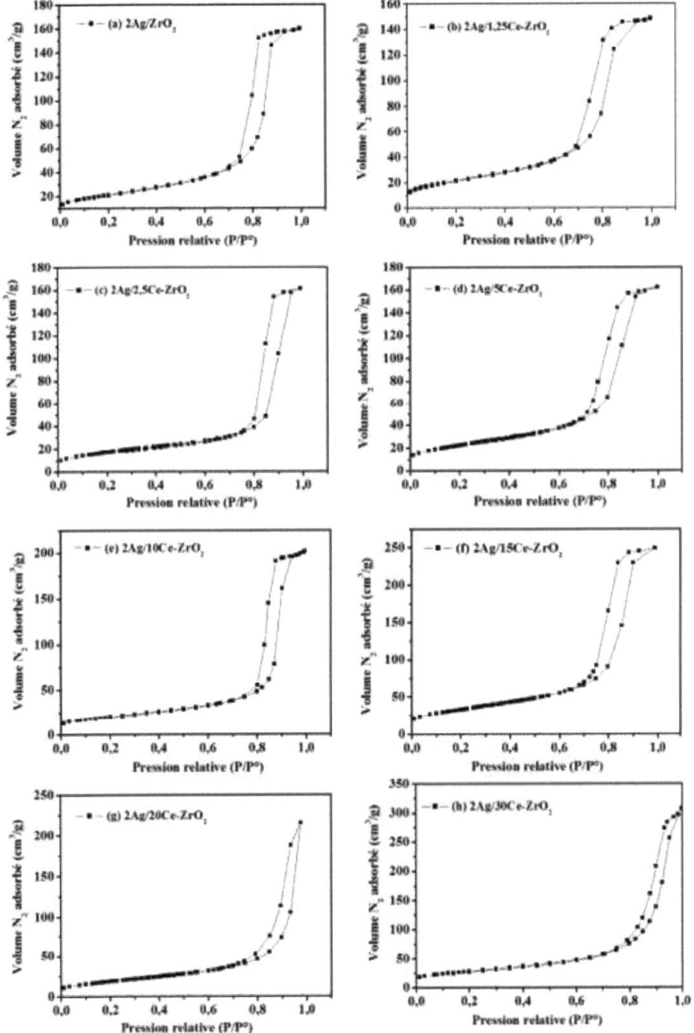

Figure 22: N2 adsorption-desorption isotherms for Ag/ZrO2 and Ag/yCe-ZrO2 catalysts.

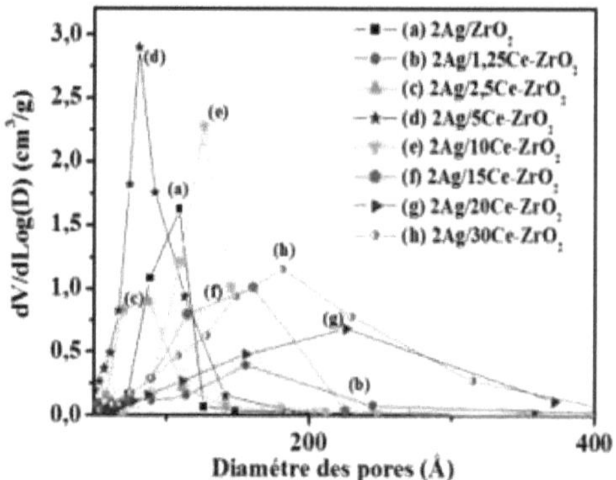

Figure 23. Pore distribution curves for Ag/ZrO2 and Ag/yCe-ZrO2 catalysts.

Analysis of the curves obtained in Figures 22 and 23 shows that the Ag/ZrO2 and Ag/yCe-ZrO2 catalysts have type IV isotherms and type H1 hysteresis loops. They are mesoporous and have cylindrical pores.

III. 3. 2. 3 UV-Visible Spectroscopy

Figure 24 shows the UV-Vis spectra of the Ag/ZrO2 and Ag/yCe-ZrO2 solids.

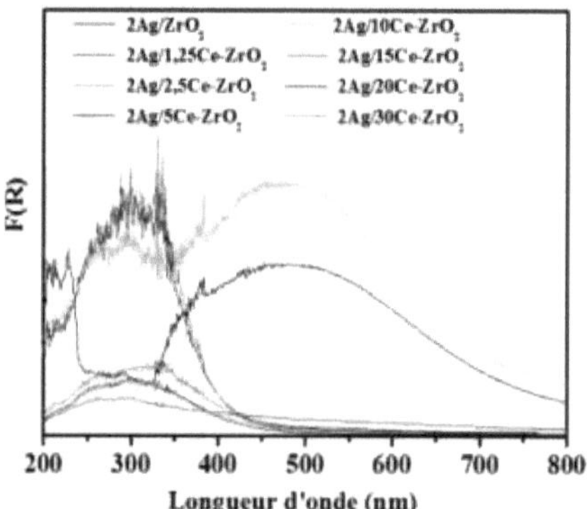

Figure 24. UV-Vis spectra of the Ag/ ZrO2 and Ag/yCe-ZrO2 catalysts.

Figure 24 shows that, in addition to the bands relating to the O^{2}-> Zr4+, O2^{-}> $Ce^{3+}$and O2-> Ce^{4+} transitions, the Ag/ZrO2 and Ag/yCe-ZrO2 catalysts exhibit bands at around 212 and 226 nm which correspond to the d-d transitions of the Ag^{+} species well dispersed on the surface of the solids [28, 30]. The bands relating to the Ag^ clusters are observed at 290 and 330 nm in the UV spectrum of the Ag/ZrO2 solid and are superimposed with the cerium absorptions in the UV spectra of the Ag/yCe-ZrO2 catalysts [25].

The band observed at 380 nm, for the Ag/ZrO2 and Ag/10Ce-ZrO2 catalysts, can be attributed to another type of Ag_n^{s} clusters+ [25,30,31] which are only formed in the case of these two samples, whereas the broad band detected at 480 nm is probably due to silver (Agn) metallic aggregates or clusters [31].

Note that the Ag/10Ce-ZrO2 sample has a different UV spectrum from the Ag/ZrO_2 and Ag/yCe-ZrO_2 samples (y^ 10), which may be linked to a change in the nature of the silver species due to the addition of Ce.

III. 3. 2. 4 X-ray photoelectron spectroscopy (XPS)

The XPS results corresponding to the Ag3d orbital in the case of the Ag/ZrO2 and Ag/10Ce-ZrO2 catalysts are shown in Table 18 and illustrated in Figure 25.

Table 18. Binding energies and oxidation state of Ag in the catalysts.

Samples	*orbitals*	*Binding energies (eV) and oxidation state*	*Ref. biblio.*
Ag/ZrO2	Ag3d5/2 Ag3d 3/2	368.1 and 374.2 eV Ag metal (Ag^0)	[7,32]
Ag/Ce-ZrO2	Ag3d5/2 Ag3d 3/2	367.6 and 374 eV Ag oxide (Ag_2O)	

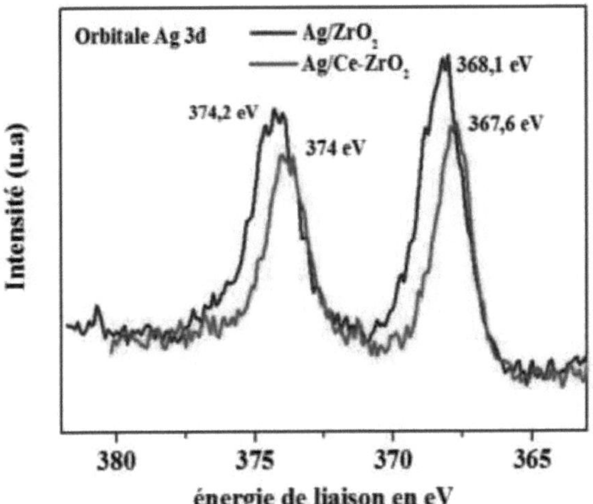

Figure 25. Ag 3d XPS spectra of the Ag/ZrO2 and Ag/Ce-ZrO2 catalysts.

Analysis of the Ag/ZrO2 and Ag/10Ce-ZrO2 samples by XPS shows that metallic silver $Ag^{(0)}$), giving peaks at 368.1 and 372.4 eV [7.30], is in the majority on the surface of the Ag/ZrO2 sample. However, oxidised silver species, with peaks at 367.6 and 374 eV [7,30], are in the majority on the surface of the Ag/10Ce-ZrO2 catalyst. These results confirm that the addition of cerium affects the nature of the supported silver species and allows them to be well dispersed on the surface of the Ag/10Ce-ZrO2 catalyst.

III. 3. 2. 5 Analysis by RTP-H2

The redox properties of the Ag/ZrO2 and Ag/yCe-ZrO2 catalysts were analysed usingRTP-H2. The results obtained are shown in Figure 26.

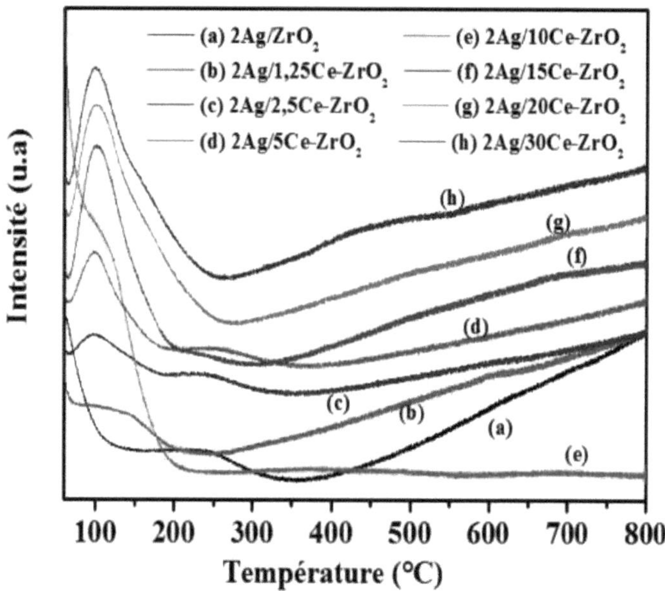

Figure 26. Results of RTP-H2 analysis of Ag/ZrO2 and Ag/yCe-ZrO2 catalysts.

The RTP-H2 curves for the Ag/yCe-ZrO2 catalysts (Figure 26) are different from those obtained for the yCe-ZrO2 supports (Figure 20) and show the appearance of new peaks at low temperatures (60-300°C). These peaks can be attributed to the reduction of silver and cerium species [25,32-34]. Z. Qu et al [25] found similar results on Ag/CeO2 catalysts and attributed the peaks between 65 and 350°C to the reduction of the different silver species (Ag^0, Ag^+) and to the oxygen associated with the silver nanoparticles. In other work [33-35] on Ag-CeO2-A, Ag/CeO2, KN-Ag/Ce-Zr and N- Ag/Ce-Zr catalysts, the peaks at low temperatures (< 300 °C) are attributed to the reduction of oxygen bound to silver or cerium. It has been reported that the presence of silver weakens the Ce-O bond associated with Ag and facilitates the reduction of cerium surface oxygen [25]. Thus, the addition of Ag influences the redox properties of Ag/yCe-ZrO2 catalysts.

III. 3. 2. 6 Analysis by DTP-NH3

The DTP-NH3 curves for some samples are shown in Figure 27.

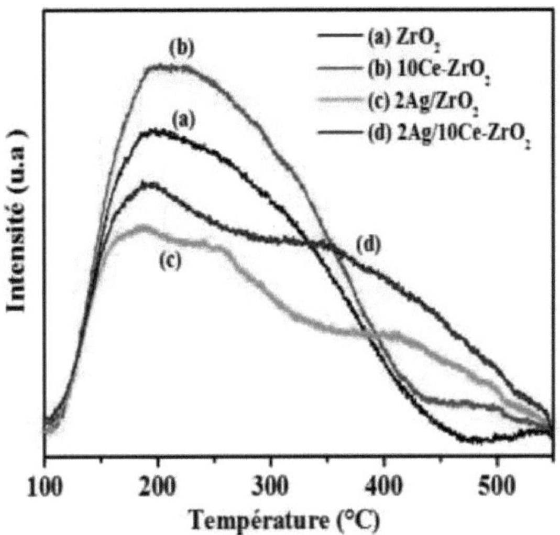

Figure 27. Courbes de DTP-NH_3 des solides.

This figure shows that the ZrO2 and 10Ce-ZrO2 catalyst supports exhibit a broad NH3 desorption peak between 100 and 400°C which can be attributed to NH3 desorption from weak acid sites (around 200°C) and medium strengths (around 300°C) [1,35-38]. The intensity of this peak is greater in the case of the Ce-ZrO2 solid, where the addition of cerium increases the total acidity of the support. The addition of Ag modifies the acidity; it leads to a decrease in the intensity of the peak located between 100 and 400°C and allows the creation of new strong acid sites (between 400 and 500°C). L. Chmielarz et al [39] also observed the formation of new strong acid sites after the addition of Ag for Mt-Al-Ag solids.

III. 3. 2. 7 Study by temperature-programmed desorption with oxygen (DTP-O2).

The nature of the oxygen species adsorbed on the surface of the samples is analysed by temperature-programmed desorption with oxygen (DTP-O2). The DTP-O2 curves obtained are shown in Figure 28.

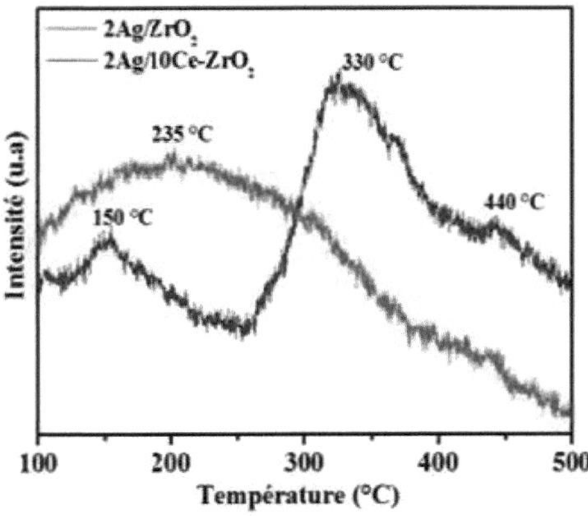

Figure 28. DTP-O2 results for the Ag/ZrO2 and Ag/Ce-ZrO2 catalysts.

The DTP-O2 curve of Ag/ZrO2 shows a broad peak with a maximum at 235°C, which can be related to molecular oxygen weakly adsorbed to the Ag surface [40,41]. The Ag/10Ce-ZrO2 sample shows three O2 desorption peaks at 150°C, 330°C and 440°C, which can be attributed to molecular oxygen weakly adsorbed on the surface, chemisorbed oxygen and oxygen linked to surface anion vacancies [39,40,42]. It should be noted that the adsorption of O2 is greater for the Ag/10Ce-ZrO2 catalyst because of the presence of cerium, which is an oxygen reservoir [43]. Previous work [15,25] has shown that the presence of a large quantity of oxygen adsorbed on the surface (O^{2-} or O^{-}) plays an important role in the stability and dispersion of silver species and improves the complete oxidation of VOCs.

III.3.3 Effect of the amount of Ag on the physicochemical properties of Ag/Ce-ZrO2 catalysts

III.3.3.1 XRD study

The X-ray diffragtograms of the xAg/10Ce-ZrO2 catalysts (with x = 0.5, 1, 2 and 3% wt.) are shown in Figure 29.

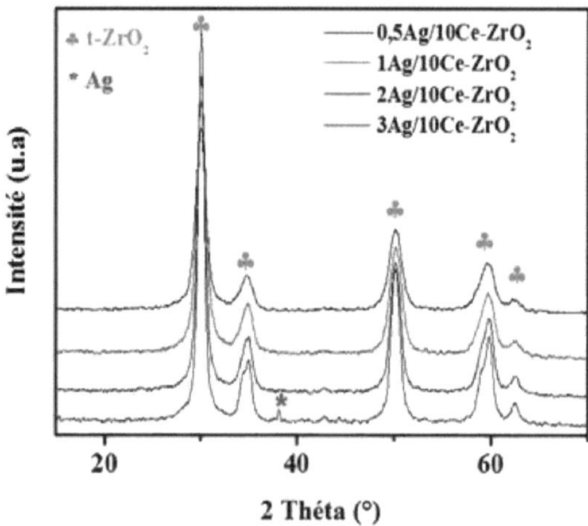

Figure 29. X-ray diffragtograms of xAg/10Ce-ZrO2 catalysts.

This figure shows the presence of peaks relating to the tetragonal phase of ZrO2. The intensity of these peaks increases as the amount of Ag increases, so the crystallinity of the support is improved by the addition of silver. The silver metal (Ag^0) lines are absent up to 2% wt Ag, while a single very weak Ag^0 peak (2θ = 38.1° (111)) appears at 3% wt Ag. Thus, we can conclude that the silver species are well dispersed on the surface of the support. K. Narasimharao et al [44] reported that Ag particles are only detected by XRD (on Ag/Fe2O3 solids) for Ag amounts greater than 10%.

III.3.3.2 Nitrogen physisorption study

The results of the analysis of xAg/10Ce-ZrO2 catalysts (with x = 0.5; 1; 2 and 3 wt.%) by nitrogen physisorption at 77 K are shown in Table 19. The N2 adsorption-desorption isotherms and pore distribution curves of the samples are shown, respectively, in Figures 30 and 31.

Table 19. Results of the analysis of xAg/10Ce-ZrO2 catalysts by nitrogen physisorption at 77 K.

Samples	*SBET (m2'-g(1))*	*Vp (cmA-gg*	*Dmoy(Å)*
0.5Ag/10Ce-ZrO2	162	0,41	71
1Ag1()Ce-Zr()2	140	0,39	121
2Ag/10Ce-ZrO2	73	0,31	127
3Ag/10Ce-ZrO2	55,8	0,25	177

The results obtained show that the xAg/10Ce-ZrO2 solids have type IV isotherms according to the IUPAC classification and type H1 hysteresis loops; they therefore have a mesoporous texture and their pores are bottle-shaped. The decrease in the specific surface area of the catalysts when the quantity of silver increases is attributed, according to the literature, to the blocking of the pores of the support by the silver species or by the agglomeration of these species on the surface [43-47]. N . Sobana, Y.Yang and R. Saravanan et al [45-47] also observed a decrease in the specific surface area of Ag/TiO2, Ag/a-Fe2O3 and Ag/ZnO catalysts, respectively, following the addition of Ag.

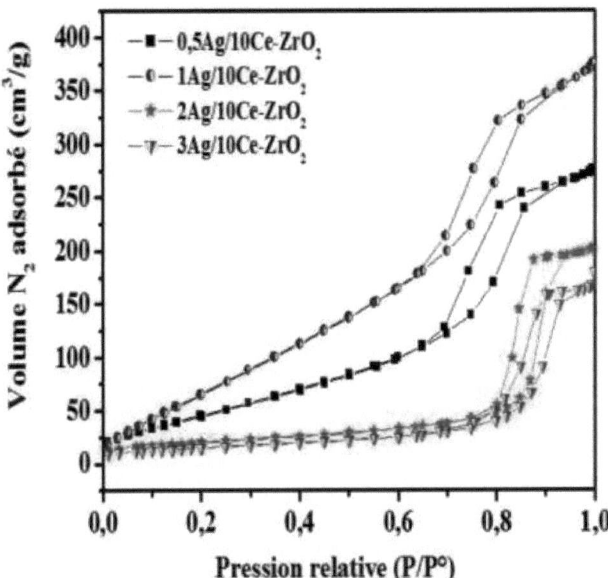

Figure 30: N2 adsorption-desorption isotherms for xAg/10Ce-ZrO2 catalysts.

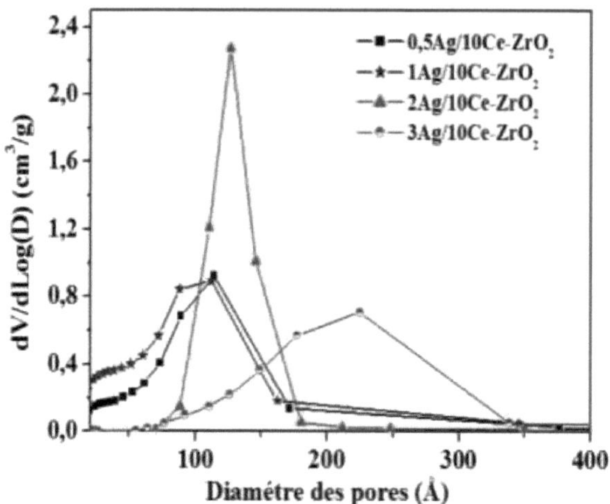

Figure 31. Pore distribution curves for xAg/10Ce-ZrO2 catalysts.

III. 3. 3. 3 UV/Visible spectroscopy in diffuse reflection

Figure 32 shows the UV-Visible diffuse reflection spectra of the xAg/10Ce-ZrO2 catalysts.

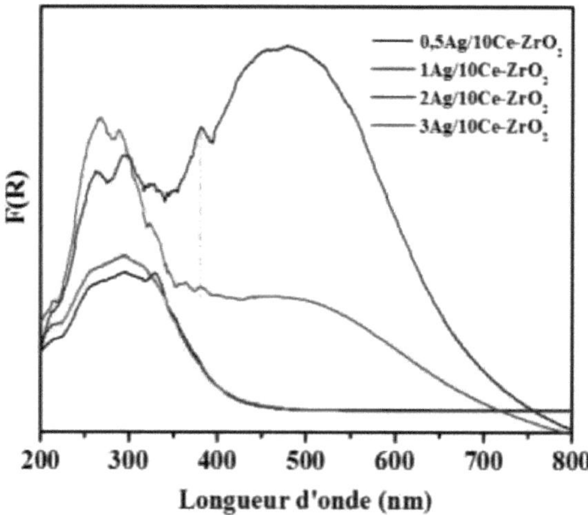

Figure 32. UV-Visible spectra of xAg/Ce-ZrO2 catalysts.

Figure 32 shows that the shape of the UV-visible spectrum varies with the variation in the amount of silver, hence the Ag content influences the nature of the Ag species formed on the surface of Ag/Ce-ZrO2 catalysts. It should also be noted that the largest quantity of Agn^{s+} and Agn clusters (observed at around 380 and 480 nm, respectively [25,28,31]) is formed on the surface of solids containing 2 % wt. Ag. They are in low quantity for x = 3 % wt Ag and absent for x = 0.5 and 1 % wt Ag.

III.4 Characterisation of yFe-ZrO2 and Ag/yFe-ZrO2 solids

III.4.1 Physicochemical properties of yFe-ZrO2 aerogel supports

III.4.1.1 XRD analysis

The X-ray diffractograms of the yFe-ZrO2 aerogels (with y = 0, 1.25, 2.5, 5 and 10% wt) are shown in Figure 33.

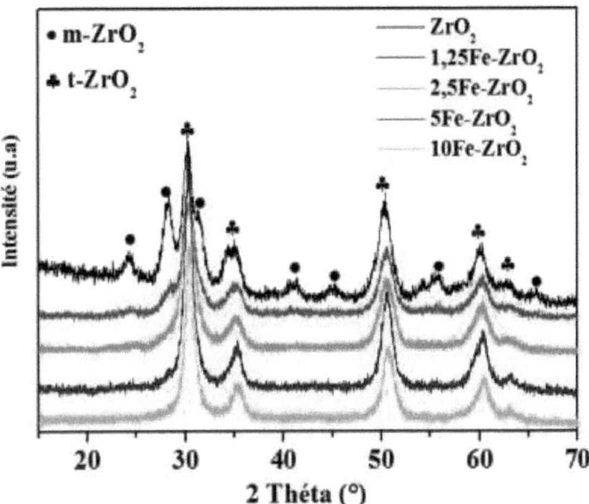

Figure 33. X-ray diffractograms of yFe-ZrO2 aerogel supports

This figure shows for yFe-ZrO2 solids present lines of the tetragonal phase of ZrO2 at 2θ = 30.33° (101), 34.67° (002), 50.37° (112), 60.09° (211), 62.57° (202) [PDF 79-1769]). The monoclinic phase of zirconia is observed at 2θ = 24.38° (110), 28.26° (-111), 31.56° (111), 41.07° (102), 45.21° (-202), 54.09° (300), 55.60 (130) is 65,68° (320) [PDF 89-9066]) for 1.25% wt Fe then disappears when the amount of Fe increases, hence the presence of iron allows stabilisation of the tetragonal phase of ZrO2. Note that the absence of Fe oxide peaks in Figure 33 reveals good dispersion of iron on the zirconia surface [48].

III. 4. 1. 2. Study by nitrogen physisorption at 77 K

The results of nitrogen physisorption at 77 K for the yFe-ZrO2 supports are shown in Figures 34 and 35 and in Table 20.

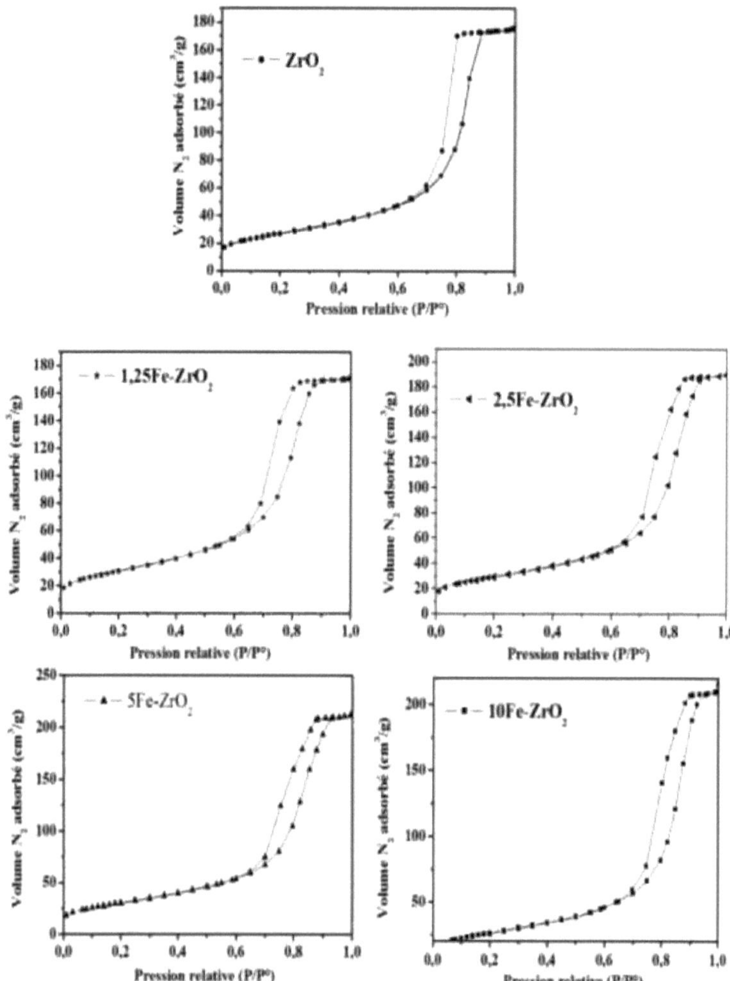

Figure 34. N2 adsorption-desorption isotherms at 77 K for yFe-ZrO2 supports

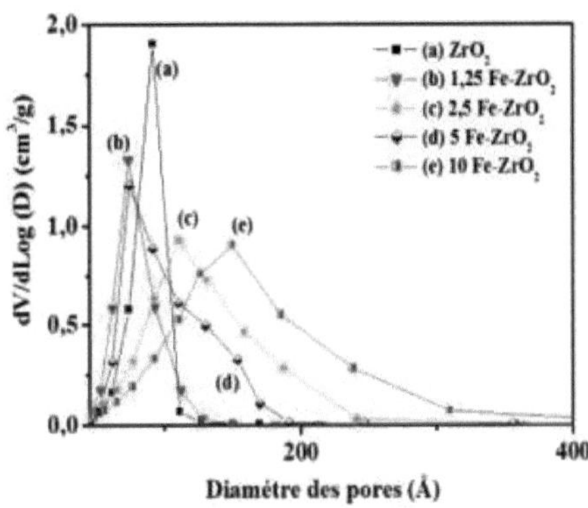

Figure 35. Pore distribution curves for yFe-ZrO2 catalyst supports.

The N2 adsorption-desorption isotherms and pore distribution curves (Figures 34 and 35) show that all the isotherms are of type IV, so the solids obtained are mesoporous with type H1 hysteresis loops showing that the pores are cylindrical.

Table 20 shows that all the samples are characterised by a large specific surface area (SBET > 93 m^2/g) which decreases slightly with increasing Fe content due to blocking of the pores of the solids by iron species or agglomeration of these species on the surface [45]. This result is similar to that obtained by Y. Zhong et al [49] on AC-Fe catalysts containing different amounts of iron.

Table 20. Results of the analysis of ZrO2 and yFe-ZrO2 aerogel supports by N2 physisorption at 77 K.

Samples	*SBET* $(m^2.gl$	*Vp (cnr.-g)*	*Dmoy*$(\AA)$
ZrO2	107	0,33	88
1.25Fe-ZrO$_2$	111	0,26	90
2,5Fe-ZrO2	107	0,29	104
5Fe-ZrO2	93	0,28	92
10Fe-ZrO$_2$	95	0,32	134

III. 4. 1. 3 Analysis by UV-Visible spectroscopy

The results of the analysis of the samples by UV-visible spectroscopy are shown in Figure 36.

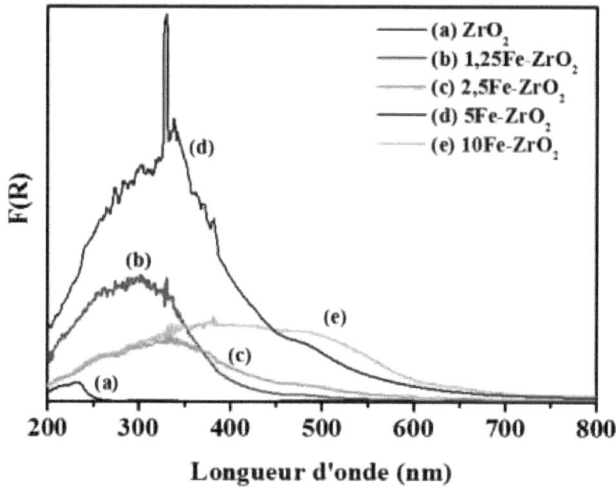

Figure 36. UV-Vis spectra of yFe-ZrO2 aerogel supports.

This figure shows the presence of several absorption bands of iron species: species in octahedral and tetrahedral coordinations for $\lambda < 300$ nm [50], FexOy agglomerates or clusters for $300 < \lambda < 400$ nm [46,51], and iron oxide aggregates (a-Fe2O3) for $\lambda > 400$ nm [46,47].
We note that Fe2O3 aggregates are only formed when the iron content exceeds 5% wt. Thus, the nature of the Fe species formed on the zirconia surface is affected by the iron content.

III. 4. 1. 4 Analysis by RTP-H2

The effect of the amount of Fe on the redox properties of the yFe-ZrO2 solids is examined by means of programmed temperature reduction using hydrogen. The results obtained are shown in Figure 37 and Table 21.

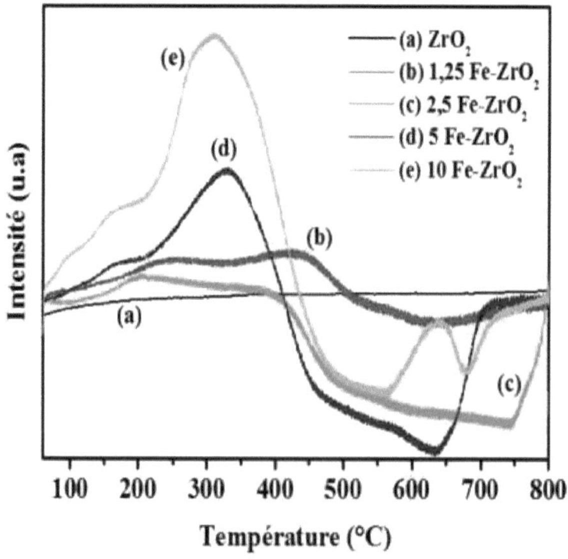

Figure 37. RTP-H2 profiles of yFe-ZrO2 aerogel supports.

Several studies [52-57] have shown that the reduction of iron species generally takes place in the following stages:

$$Fe_2O_3 \rightarrow Fe_3O_4 \rightarrow FeO \rightarrow Fe$$

$$3Fe_2O_3 + H_2 \rightarrow 2Fe_3O_4 + H_2O \ (1)$$

$$Fe_3O_4 + H_2 \rightarrow 3FeO + H_2O \ (2)$$

$$FeO + H_2 \rightarrow Fe + H_2O \ (3)$$

The RTP-H2 profiles of the yFe-ZrO2 catalyst supports (Fig. 37) show several reduction peaks at low temperatures which can be attributed to the reduction of different iron species (see Table 21). It is likely that the nature and interactions of these species with zirconia are affected by the iron content. F.G.E. Nogueira et al [57] studied the redox properties of iron supported on montmorillonite and reported that the interaction between Fe and the support influences the nature of the iron species.

Table 21. Results of RTP-H2 analysis of yFe-ZrO2 supports.

Samples	*Peak temperature*	*Allocation to corresponding reductions*	W. *Biblio.*
1.25Fe-ZrO$_2$	240 °C and 447 °C	Fe2Ü3^ FesÜ4	[53,58]
2,5Fe-ZrO$_2$	194°C and 397°C	Fe2Ü3^Fe3O4	[51,59]
5Fe-ZrO$_2$	164 °C and 328 °C	Fe2Ü3^Fe3Ü4	[49]
10Fe-ZrO2	91 °C, 160 °C and 309 °C	Fe2Ü3 ^ Fe3Ü4	[48-49]
	638 °C	Fe$_3$Ü4 > FeO ^ Fe clusters de Fe$_3$Ü4 > Fe^0	[49]

These results show that the addition of iron leads to the formation of redox sites on the surface of ZrO2.

III. 4. 1. 5 Analysis by DTP-NH3

The DTP-NH3 curves obtained on the ZrO2 and yFe-ZrO2 solids are shown in Figure 38.

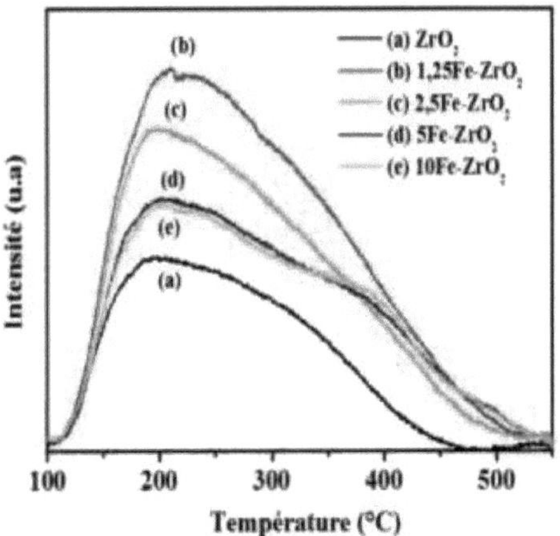

Figure 38. DTP-NH3 profiles of ZrO2 and yFe-ZrO2 aerogel supports.

Analysis of this figure shows that the total acidity of zirconia increases with the addition of iron. Indeed, the highest acidity is obtained for 1.25 wt% Fe [34-38], then it decreases with the increase in the quantity of iron. This may be due to interactions between iron species and zirconia which affect the acidic properties of yFe-ZrO2 aerogels.

The modification of the acidity of various solids (Fe-Mn-Ce/TiO2, Fe-MCM-41, Fe-ZSM-5, *FexCuxSiBEA* and *FexCuxHAlBEA*) by the addition of Fe has been reported in many previous works [60-64].

III.4.2 Characterisation of Ag/ZrO2 and Ag/yFe-ZrO2 catalysts

III.4.2.1 XRD study

The X-ray diffractograms of the Ag/yFe-ZrO2 catalysts (Figure 39) are similar to those of the yFe-ZrO2 supports (Figure 33), so the presence of silver does not affect the structure of the catalysts. It is important to note that the absence of peaks relating to metallic silver and Ag oxides for the Ag/yFe-ZrO2 catalysts reveals a good dispersion of silver species, which is probably due to interactions between silver and iron. A weak peak for metallic silver is observed at 2θ = 38, 1° (111) (PDF 65-2871)) on the X-ray diffractogram of the Ag/ZrO2 solid.

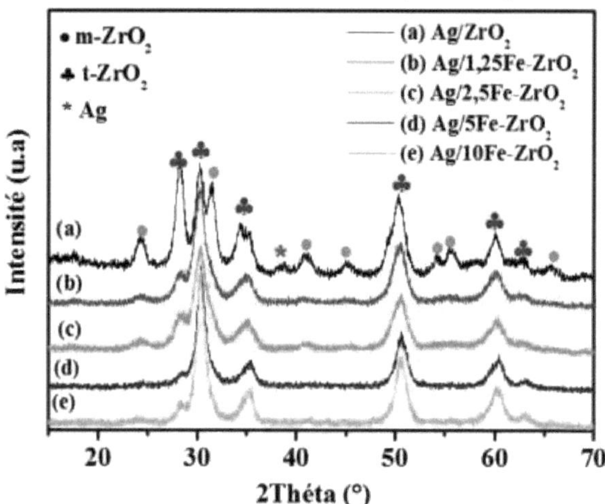

Figure 39. X-ray diffragtograms of Ag/yFe-ZrO2 catalysts.

III.4.2.2 Analysis by nitrogen physisorption at 77 K

The N2 adsorption-desorption isotherms and the pore distribution curves for the Ag/yFe-ZrO2 catalysts are shown in Figures 40 and 41 respectively.

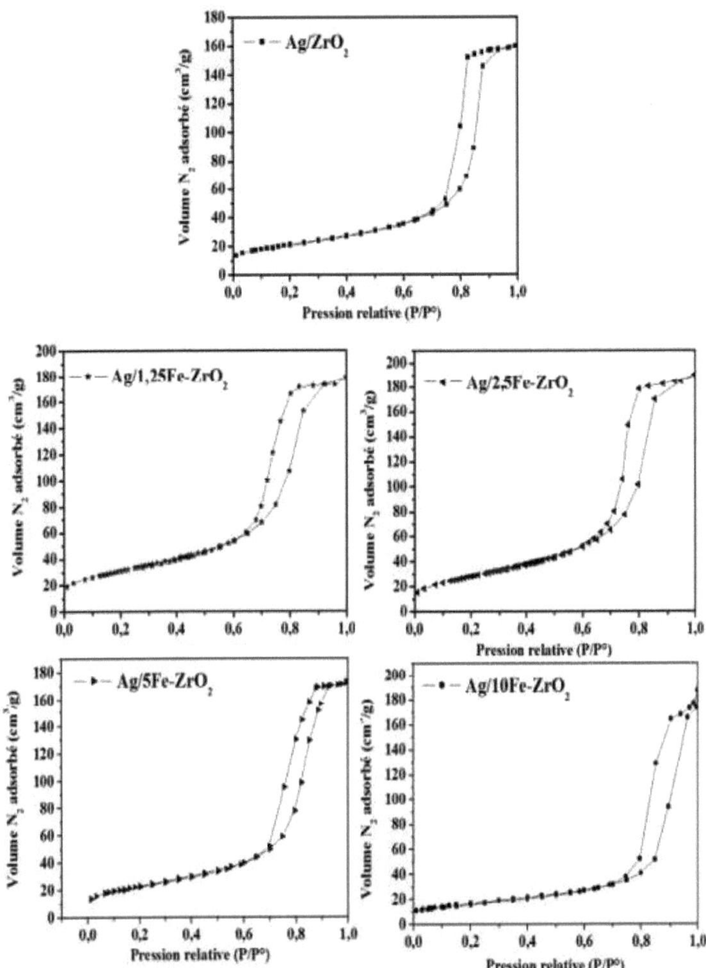

Figure 40. N2 adsorption-desorption isotherms for Ag/yFe-ZrO2 catalysts.

The Ag/yFe-ZrO2 solids show type IV isotherms (Figure 40) and type H1 hysteresis loops (Figure 40). They are mesoporous with cylindrical pores.

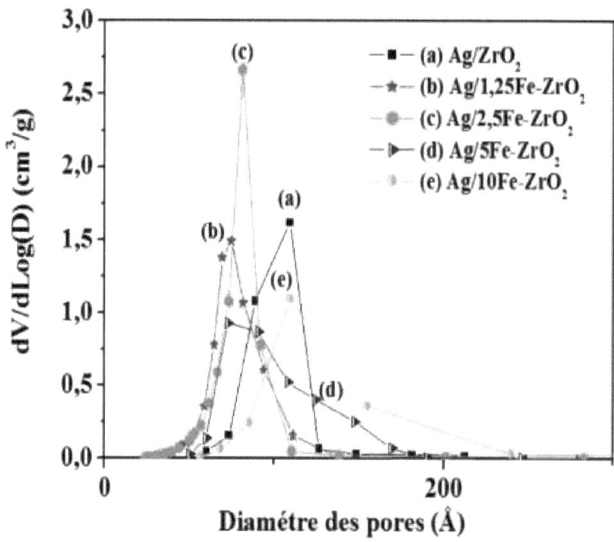

Figure 41. Pore distribution curves for Ag/ZrO2 and Ag/yFe-ZrO2 catalysts.

The textural properties of Ag/yFe-ZrO2 materials are shown in Table 22.

Table 22. Results of the analysis of 2Ag/yFe-ZrO2 catalysts by N2 physisorption.

Samples	*SBET (m².gl*	*Vp (cmA-gg*	*Dmoy(Å)*
2AgZrO$_2$	76	0,24	97
2Ag1,25Fe-ZrO$_2$	112	0,27	84
2Ag2,5Fe-ZrO$_2$	105	0,29	87
2Ag5Fe-ZrO$_2$	84	0,26	116
2Ag/10Fe-ZrO2	59	0,26	161

Comparison of the values given in this table with those in Table 20 shows that the addition of silver does not affect the textural properties of the aerogel supports for a quantity of iron< 2.5 % wt. However, a decrease in the specific surface area is observed with the addition of a quantity of iron> 5 % wt. This is probably due to the blocking of the pores of the solids by the iron and silver species.

III. 4. 2. 3 Study by UV-Visible spectroscopy

The UV-visible spectra of the Ag/ZrO2 and Ag/yFe-ZrO2 catalysts are shown in Figure 42.

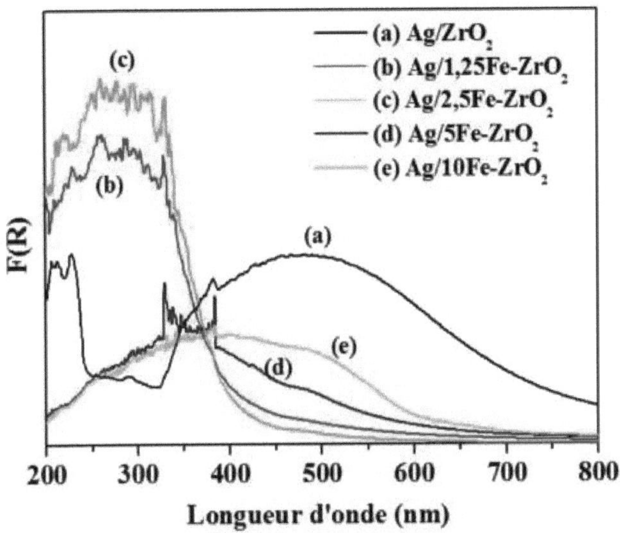

Figure 42. UV-Vis spectra of the Ag/ZrO2 and Ag/yFe-ZrO2 catalysts.

Based on this figure and Table 23, we can suggest the presence (in all catalysts) of the following species: Fe3+, Ag+ and $Ag^{s}{}_{+n}$; which absorb at $\lambda < 300$ nm [46], 210 nm < λ <240 nm [28,30,31] and 290 nm < λ < 390 nm [25,28,31], respectively. Ag^{s} clusters$_{+n}$ (which absorb at 390 nm [31]) and Ag_n (which absorb at 480 nm [31,65]), FexOy clusters and Fe2O3 iron oxide aggregates (which absorb for $\lambda >$ 390 nm) are present only in the case of Ag/5Fe-ZrO2 and Ag/10Fe-ZrO2 samples.

Table 23. UV-vis bands for Zr, Fe and Ag species according to the literature.

Species	*Nature of the absorption band*	*Wavelengths*	*References*
Zr	2 22 /z-\2- 4+ 4+\ charge transfer (O ^ Zr)	200 and 250 nm	[7,8]
Fe	Pseudo-tetrahedral Fe (III)	247 and 845 nm	[66]
	Fe<$O^{(2-)}Fe^{3+}$ ions isolated in octahedral and tetrahedral coordination	< 300 nm	[46]
	iron oxide aggregates (a-Fe_2O_3)	440-550 nm	[46-47]
	iron oxide aggregates (a-Fe_2O_3)	>400 nm	[46-47]
Ag	d-d transition ($4d^{10}$-$4d^{9}{}_{5s1}$) of Ag^+	210-240 nm	[28,30,31]
	$Ag^{\wedge}{}_n$ clusters	275 nm	[25,28,31]
	Aggregates or metal clusters of Ag	350 and 450 nm	[31,61]

III. 4. 2. 4 Analysis by RTP-H2

The RTP-H2 curves obtained on the Ag/yFe-ZrO2 catalysts (where: y = 0; 1.25 ; 2.5; 5 and 10% wt.) are shown in Figure 43.

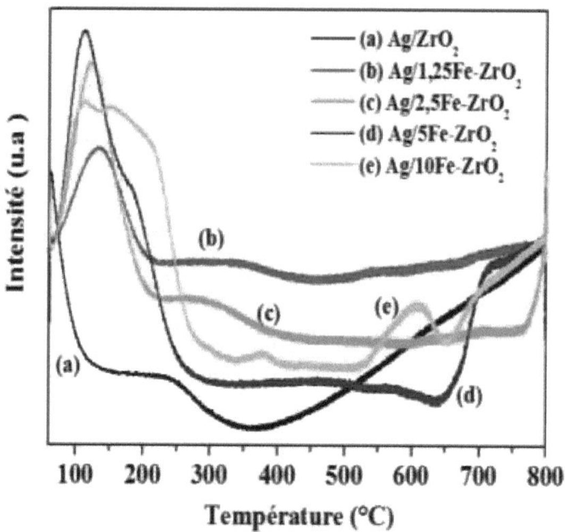

Figure 43. Results of RTP-H2 on Ag/ZrO2 and Ag/yFe-ZrO2 catalysts.

The RTP-H2 profiles of Ag/yFe-ZrO2 catalysts show peaks between 50 and 300°C which can be attributed to the reduction of silver and oxidised iron species [25,28,67,68]. Comparison of these results with those for yFe-ZrO2 supports (fig 37) shows that iron reduction occurs at lower temperatures in the presence of Ag. This may be due interactions between the silver and iron species [63]. X. Zhang et al [68] have reported that the addition of Ag weakens the Fe-O bonds in Ag/Fe2O3 because there are interactions between oxygen and silver which allow reduction of iron at lower temperatures.

III. 4. 2. 5 DTP-NH3 study

The DTP-NH3 curves for the Ag/yFe-ZrO2 catalysts are shown in Figure 44.

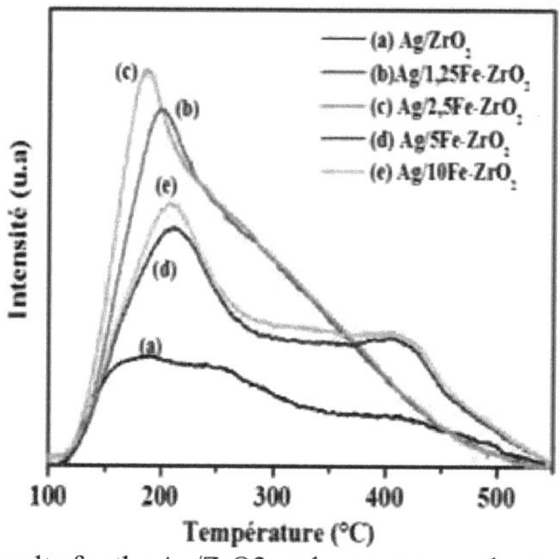

Figure 44. DTP-NH3 results for the Ag/ZrO2 and Ag/yFe-ZrO2 catalysts

Comparison of these results with those for yFe-ZrO2 solids (Fig. 38) shows that new strong acid sites are created on the surface of Ag/5Fe-ZrO2 and Ag/10Fe-ZrO2 catalysts (between 400 and 500°C) after the addition of Ag. Thus, the presence of silver species affects the acidity of the Ag/yFe-ZrO2 catalysts.

III.4.3 Effect of Ag content on the physicochemical properties of xAg/5Fe-ZrO2 catalysts

III.4.3.1 XRD analysis

XRD diffractograms of the xAg/5Fe-ZrO2 catalysts (x= 0.5, 2, 2.5 and 3% wt.) are shown in Figure 45.

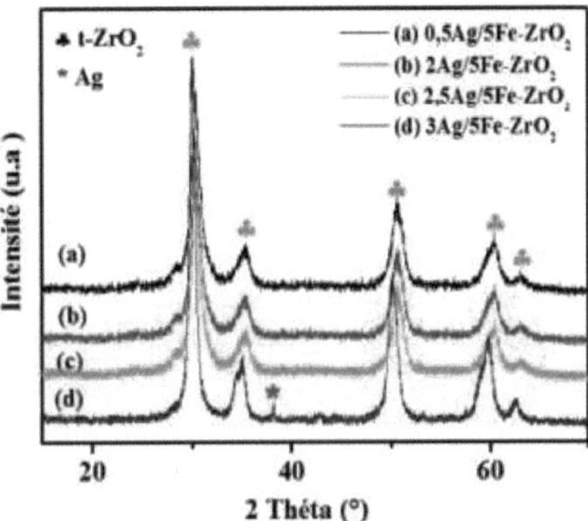

Figure 45. X-ray diffractograms of xAg/5Fe-ZrO2 catalysts.

The xAg/5Fe-ZrO2 catalysts show tetragonal phase lines and 2θ = 30.33° (101), 34.67° (002), 50.37° (112), 60.09° (211), 62,57° (202) [PDF 79-1769] and the monoclinic phase at 2θ = 24.38° (110), 28.26° (-111), 31.56° (111), 41.07° (102), 45.21° (202), 54.09° (300), 55.60 (130) and 65.68° (320) [PDF 89-9066]. These no longer exist for a quantity of Ag = 3% wt, which confirms the stabilisation of the tetragonal phase of ZrO2 by the presence of Ag. A weak peak is observed at 2θ = 38.1° (111) (PDF 65-2871) for the 3Ag/5Fe-ZrO2 solid, due to the formation of a very small amount of metallic silver.

III.4.3.2 Nitrogen physisorption study

The N2 adsorption-desorption isotherms and the pore distribution curves for the xAg/5Fe-ZrO2 catalysts (Figures 46 and 47) show that all the isotherms are type IV and the hysteresis loops are type H1.

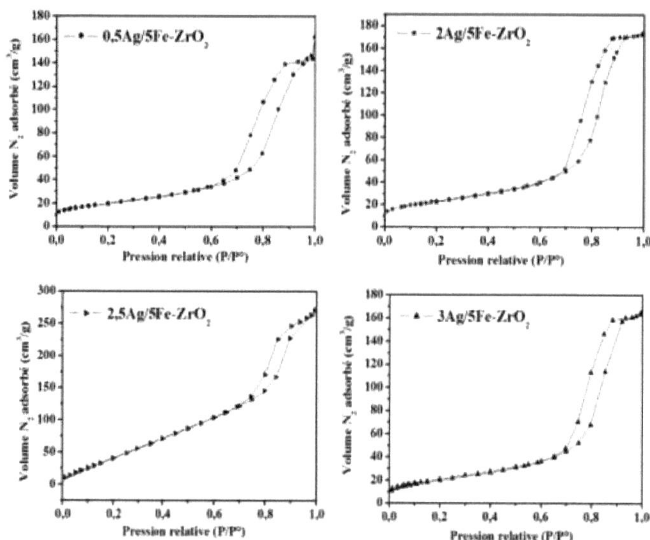

Figure 46: N2 adsorption-desorption isotherms for xAg/5Fe-ZrO2 catalysts

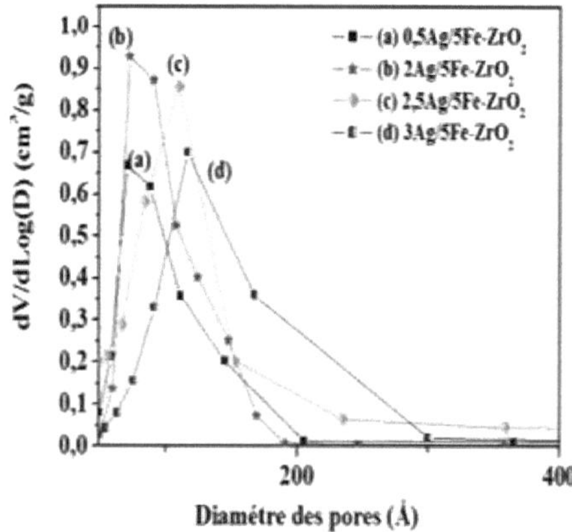

Figure 47. Pore distribution curves for xAg/5Fe-ZrO2 catalysts.

The results in Table 24 show that the xAg/5Fe-ZrO2 catalysts (x= 0.5; 2; 2.5 and 3 % wt.) are characterised by a large specific surface area (SBET> 70 cm^2/g) and a large porosity (Vp> 0.25 cm^3/g).

Table 24. Results of the analysis of xAg/5Fe-ZrO2 catalysts by N2

Samples	SBET (m^2.g'1	Vp (cnr.-g)	Dmoy(Å)
0.5Ag/5Fe-ZrO2	70	0,25	119
2Ag5Fe-ZrO_2	84	0,26	116
2,5Ag5Fe-ZrO_2	160	0,40	66
3Ag/5Fe-ZrO2	74	0,25	119

III.4.3.3 Study by UV-Visible spectroscopy

The results of the analysis of the xAg/yFe-ZrO2 catalysts by UV-visible spectroscopy are shown in Figure 48.

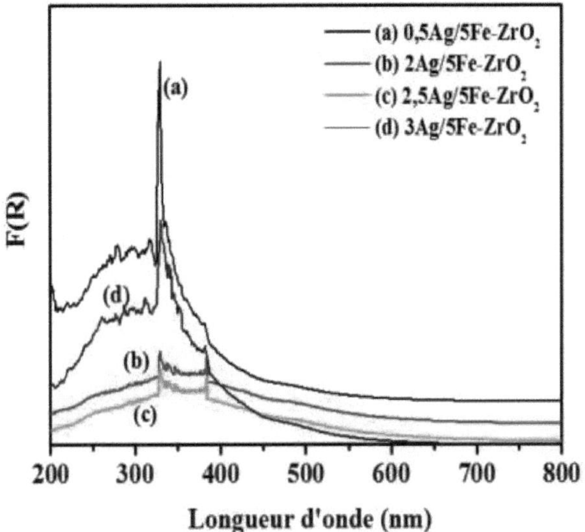

Figure 48. UV-Vis spectra of xAg/5Fe-ZrO2 catalysts.

Figure 48 shows that the catalysts exhibit absorption bands between 200 and 500 nm, revealing the presence of different silver and iron species on the surface of the xAg/5Fe-ZrO2 catalysts (Table 25).

Table 25. UV-Visible bands observed on the spectra of xAg/5Fe-ZrO2 solids.

Samples	*Wavelengths (nm)*	*Band allocation*	*Ref*
xAg/5Fe-ZrO2	< 300 nm 210-240 nm	$Fe<O^{2-}$ of Fe^{3+} ions isolated in octahedral and tetrahedral coordination d-d transition ($4d^{10}$-$4d^{9}5s1$) of Ag^{+}	[46] [28,30,31]
	290, 330 and 380 nm	Different Ag aggregates or clusters$°^{+}_{n}$	[25,28,31]

III. 4. 3. 4 RTP-H2 study

The study of the effect of Ag content on the redox properties of xAg/5Fe- ZrO2 catalysts leads to the results shown in Figure 49.

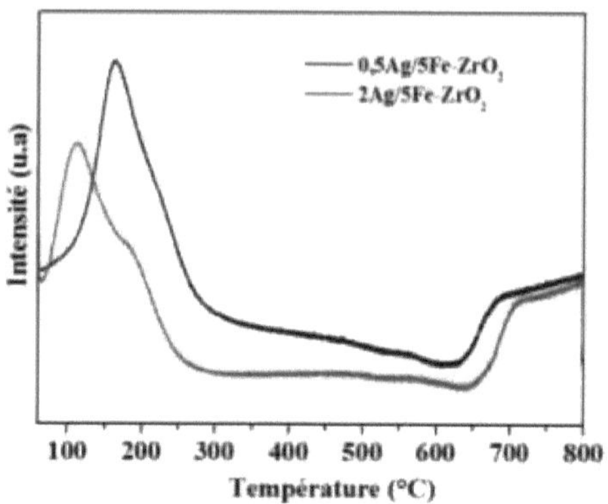

Figure 49: RTP-H2 results for xAg/5Fe-ZrO2 solids.

Analysis of these curves shows that the reduction of iron and silver is influenced by the amount of Ag: it takes place at lower temperatures when the amount of Ag increases. This is probably due to the interactions between iron and silver, K. Narasimharao et al [44] reported similar results which showed a decrease in the reduction temperature of iron by the addition of Ag to an Ag-Fe2O3 system.

111.5. Conclusion

The results obtained in this chapter show that :

* Zirconia prepared by the Sol-Gel method is characterised by a structure containing the stable monoclinic and metastable tetragonal phase of ZrO2, a high specific surface area and a high pore volume (SBET = 107 $m^2.g^{-1}$ and Vp = 0.33 $cm^3.g^{-1}$).
* Addition of M (M= Ce or Fe):
- Stabilises the tetragonal phase of zirconia.
- Leads to M-ZrO2 aerogels characterised by : SBET > 72 $m^2.g^{-1}$ and Vp > 0.21 $cm^3.g^{-1}$
- Allows the formation of acid and redox sites on the surface of *M-ZrO2* solids

* Increasing the M content has the following effects:
- Reduce the specific surface area of M-ZrO2 due to the blocking of certain pores by M species.
- Influencing the nature and dispersal of M. species.
- Modifying the acid and redox properties of *M-ZrO2* catalyst supports

* The addition of increasing amounts of Ag leads to Ag/yM-ZrO2 catalysts with a tetragonal ZrO2 phase structure, a specific SBET surface area > 59 $m^2.g^{-1}$ and a pore volume Vp > 0.22 $cm^3.g^{-1}$. It also enables:
- The formation of silver species on the surface of yM-ZrO2 aerogels is affected by the nature and quantity of M and by the Ag content.
- A transformation of the surface acidity and an improvement in the redox characteristics of the Ag/yM-ZrO2 catalysts, most probably due to the interactions between Ag and M.

III. 6. References

1. J. Arfaoui, A. Ghorbel, C. Petitto, G. Delahay, *Novel V_2O_5-CeO_2-TiO_2-SO_4^{2-} nanostructured aerogel catalyst for the low temperature selective catalytic reduction of NO by NH3 in excess O2*, Appl Catal B Environ. 224 **(2018)** 264-275.
2. C. Gannoun, R. Delaigle, D. P. Debecker, P. Eloy, G. Ghorbel, E. M. Gaigneaux, *Solgel*

derived V2O5-TO2 mesoporous materials as catalysts for the total oxidation of chlorobenzene, Catal. Comm. 15 **(2011)** 1-5.
3. Rescoll technology centre, technical file on the Sol-Gel process.
4. Dr. Ir. François Collignon, CER TECH, Sol-Gel: soft chemistry for innovation **(2011).**
5. W. Khaodee, N. Tangchupong, B. Jongsomjit, P. Praserthdam and S. Assabumrungrat. *A study on isosynthesis via CO hydrogenation over ZrO2-CeO$_2$ mixed oxide catalysts.* Catal Commun. 10 **(2009)** 494-501.
6. V. Solinas, E. Rombi, I. Ferino, M. G. Cutrufello, G. Colon and J. A. Navio. *Preparation, characterisation and activity of CeO2-ZrO2 catalysts for alcohol dehydration.* J Mol Catal A Chem. 204-205 **(2003)** 629-635.
7. Z. Guerra-Que , G. Torres-Torres, H. Pérez-Vidal, I. Cuauhtémoc-Lopez, A. Espinosa de los Monteros, J. N. Beltramini, D. M. Frias-M'arquez, *Silver nanoparticles supported on zirconia-ceria for the catalytic wet air oxidation of methyl tert-butyl ether.* RSC Adv. 7 **(2017)** 3599-3610.
8. S. Mallick, S. Rana and K. Parida, *A facile method for the synthesis of coppermodified amine-functionalized mesoporous zirconia and its catalytic evaluation in C-S coupling reaction*, Dalton Trans. 40 **(2011)** 9169-9175.
9. E. G. Heckert, A. S. Karakoti, S. Seal, W.T. Self, *The role of cerium redox state in the SOD mimeticactivity of nanoceria*, Biomaterials 29 **(2008)** 2705-2709.
10. G. Rao. and H. Zahu, *XRD and UV-Vis diffuse reflectance analysis of CeO2-ZrO2 solid solutions synthesized by combustion method*, Proc. Indian Acad. Sci. (Chem. Sci.) 11 **(2001)** 651-658.
11. C. Gannoun, R. Delaigle, D.P. Debecker, P. Eloy, G. Ghorbel, E.M. Gaigneaux, *Effect of support on V2O5 catalytic activity in chlorobenzene oxidation*, Appl. Catal. A: General 447-448 **(2012)** 1-6.
12. S. Ardizzone, M. G. Cattania and P. Lugo: *Interfacial electrostatic behaviour of oxides: Correlations with structural and surface paramaters of the phase*, Electrochim. Acta.39 (11/12) **(1994)** 1509-1517.
13. S. Ardizzone, C. L. Bianchi, M. Signoretto, *Zr_(IV) surface chemical state and acid features of sulphated-zirconia samples*, Appl. Surf. Sci. 136 **(1998)** 213-220.
14. M. Piumetti, S. Bensaid, N. Russo, D. Fino. *Nanostructured ceria-based catalysts for soot combustion: investigations on the surface sensitivity.* Appl Catal B Environ. 165 **(2015)** 742-751.
15. S. Wu, Y. Yang, C. Lu, Y. Ma, S. Yuan, G. Qian, *Soot oxidation over CeO2 or Ag/CeO2. influences of bulk oxygen vacancies and surface oxygen vacancies on activity and stability of catalyst.* **(2018).** https://doi.org/10.1002/ejic.201800423
16. X. Wu, L. Xu, D. Weng .*The thermal stability and catalytic performance of Ce-Zr promoted Rh-Pd/g-Al2O3 automotive catalysts.* Appl Surf Sci. 221 **(2004)** 375-383.
17. X. Deng, M. Li, J. Zhang, X. Hu, J. Zheng, N. Zhang, B. H. Chen, *Constructing nanostructure on silver/ceria-zirconia towards highly active and stable catalyst for soot oxidation*, Chem. Eng. J 313 **(2017)** 544-555.
18. Z. Yan'e, X. Zhu, H. Yang, W. Zua, K. Li, Y. Wang ,Y. Wei . *Characteristic of macroporous CeO2-ZrO2 oxygen carrier for chemical-looping steam methane reforming.* J Rare Earth. 32(9) **(2014)** 842-848.
19. S. Aouad, *Etude des propriétés physico-chimiques et catalytiques des solides à base de ruthénium : applications dans des réactions d'intérêts environnementaux : oxydation des*

suies, des COV et vaporeformage du méthanol, PhD thesis **(2007)** Université du Littoral Côte d'Opale, France.
20. E. Abi-Aad, *Préparations, caractérisations et aspects catalytiques d'oxydes ternaires à base de cuivre, aluminium et cérium,* Doctoral thesis **(1996)** Université du Littoral Côte d'Opale, France.
21. S. Liu, X.D. Wu, W. Liu, W.M. Chen, R. Ran, M. Li, D. Weng, *Soot oxidation over CeO2 and Ag/CeO2: Factors determining the catalyst activity and stability during reaction*, J. Catal. 337 **(2016)** 188-198.
22. H.C. Yao, and Y.F.Y. Yao, *Ceria in automotive exhaust catalysts: I. Oxygen storage*, J. Catal. 86 **(1984)** 254-265.
23. F. Ying F, S. J Wang, C. T Au and S.Y. Lai, *Highly active and stable mesoporous Au/CeO2 catalysts prepared from MCM-48 hard-template*, Microporous Mesoporous Mater. 42 **(2011)** 308-315.
24. S. mamura, H. Yamada, K. Utani, *Combustion activity of Ag/CeO2 composite catalyst*, Appl. Catal. A: General, 192 **(2000)** 221-226.
25. Z. Qu , F. Yu, X. Zhang, Y. Wang, J. Gao. *Support effects on the structure and catalytic activity of mesoporous Ag/CeO2 catalysts for CO oxidation*. Chem Eng J 229 **(2013)** 522-532.
26. J. Arfaoui, L. K Boudali, A. Ghorbel, G. Delahay . *Influence of the nature of titanium source and of vanadia content on the properties of titanium-pillared montmorillonite.* J Phys Chem Sol. 69 **(2008)** 1121-1124.
27. W. H. Shen, X. P. Dong, Y. Zhu, H. R Chen, J. L. Shi, *Mesoporous CeO2 and CuO loaded mesoporous CeO2: synthesis, characterization, and CO catalytic oxidation property,* Microporous Mesoporous Mater. 85 **(2005)** 157-162.
28. L. Ma, D. Wang ,J. Li, B. Bai , L. Fu, Y. Li, *Ag/CeO2 nanospheres: efficient catalysts for formaldehyde oxidation*, Appl. Catal. B: Environ 148-149 **(2014)** 36-43.
29. J. Arfaoui, *Synthèse et caractérisation de Montmorillonites pontées au Titane et Dopées au Vanadium : Application catalytique à la réduction sélective de NO par NH3 et à l'époxydation des alcools allyliques*, PhD thesis (**2010),** Faculté des Sciences de Tunis, Tunisie.
30. K. A Bethke, H. H Kung, *Supported Ag Catalysts for the Lean Reduction of NO with C3H6*. J Catal. 172 **(1997)** 93-102.
31. L. Kundakovic, M. Flytzani-Stephanopoulos, *Deep oxidation of methane over zirconia supported Ag catalysts,* Appl. Catal. A: Gen. 183 **(1999)** 35-51.
32. J. Zhang, L. Li, X. Huang, G. Li. *Fabrication of Ag-CeO2 core-shell nanospheres with enhanced catalytic performance due to strengthening of the interfacial interactions*. J Mater Chem. 22 (**2012)** 10480-10487.
33. S. Scire, P. M Riccobene, C. Crisafulli, *Ceria supported group IB metal catalysts for the combustion of volatile organic compounds and the preferential oxidation of CO*, Appl. Catal. B: Environ. 101 **(2010)** 109-117.
34. X. Deng, M. Li, J. Zhang, X. Hu, J. Zheng, N. Zhang, B. H. Chen, *Constructing nanostructure on silver/ceria-zirconia towards highly active and stable catalyst for soot oxidation*, **(2016)** http://dx.doi.org/10.1016/j.cej.2016.12.088.
35. C. Petitto H. P. Mutin, G. Delahay. *Hydrothermal activation of silver supported alumina catalysts prepared by sol-gel method: Application to the selective catalytic reduction (SCR) of NOx by n-decane*. Appl. Catal. B. 134-135 **(2013)** 258-264.
36. J. Arfaoui, K. L Boudali, A. Ghorbel. *Catalytic epoxidation of allylic alcohol (E)-2-*

Hexen-1-ol over vanadium supported on unsulfated and sulfated titanium pillared montmorillonite catalysts: Effect of sulfate groups and vanadium loading. Appl Clay Sci. 48 **(2010)** 171-178.
37. J. Arfaoui, K. L. Boudali, A. Ghorbel, G. Delahay, *Characterization and catalytic performance of vanadium supported on sulfated Ti-PILC catalysts issued from different Ti-precursors in selective catalytic reduction of nitrogen oxide by ammonia*. J Mater Sci, 44 **(2009)** 6670-6676.
38. J. Arfaoui, K. L. Boudali, A. Ghorbel , D. Delahay. *Effect of vanadium on the behaviour of unsulfated and sulfated Ti-pillared clay catalysts in the SCR of NO by* NH_3. Catal Today. 142 **(2009)** 234-238
39. L. Chmielarz, M. Zbroja, P. Kuretrowski, B. Dudek, A. Rafalska-£asocha and R. Dziembaj, *PILLARED MONTMORILLONITES MODIFIED WITH SILVER Temperature programmed desorption studies*, Journal of Thermal Analysis and Calorimetry, 77 **(2004)** 115-123.
40. S. Z. Todorova, H. G. Kolev, M. G. Shopska, G. B. Kadinov, J. P. Holgado, A. Caballero. *Silver-based catalysts for preferential CO oxidation in hydrogen-rich gases (PROX)*. Bulg Chem Commun. 50 **(2018)** 17-23.
41. Y. E Sung, W.Y. Lee, H. K Rhee, H. I Lee, *The effect of oxygen on the chemisorption on polycrystalline silver surface*. Kor J Chem Eng. 6 **(1989)** 300-305
42. L. Kongzhai, W. Hua , W. Yonggang, L. Mingchun. *Catalytic performance of cerium iron complex oxides for partial oxidation of methane to synthesis gas*, J. Rare EARTH. 26 **(2008)** 705-710.
43. T. Tabakova, T. F. Boccuzzi, M. Manzoli, J.W. Sobczak, V. Idakiev, D. Andreeva, *A comparative study of nanosized IB/ceria catalysts for low-temperature water-gas shift reaction*, Appl. Catal. A 298 **(2006)** 127-143.
44. K. Narasimharao, A. Al-Shehri, S.Al-Thabaiti, *Porous Ag Fe2O3 nanocomposite catalysts for the oxidation of carbon monoxide*, Appl. Catal. A: Gen , 505 (**2015**) 431440.
45. N. Sobana, M. Muruganadham, M. Swaminathan, *Nano-Ag particles doped with TiO2 for efficient photodegradation of Direct azo dyes*, J. Mol. Catal. A: Chem. 258 **(2006)** 124-132.
46. Y. Wang, Y. Wang, J. Cao, F. Kong, H. Xia, J. Zhang, B. Zhu, S. Wang, S. Wu, *Low-temperature* H_2S *sensors based on Ag-doped -Fe2O3 nanoparticles*, Sensors and Actuators B 131 **(2008)** 183-189.
47. R. Saravanan, N. Karthikeyan , V.K. Gupta, E. Thirumal, P. Thangadurai, V. Narayanan, A. Stephen, *ZnO/Ag nanocomposite: An efficient catalyst for degradation studies of textile effluents under visible light*, Materials Science and Engineering C 33 **(2013)**2235-2244.
48. H. Y. Chen , W. M.H. Sachtler, *Activity and durability of Fe/ZSM-5 catalysts for lean burn NOx reduction in the presence of water vapor,* Catal. Today 42 **(1998)** 73-83.
49. Y. Zhong, G. Li , L. Zhu, Y,Yan, G. Wu, C. Hu, *Low temperature hydroxylation of benzene to phenol by hydrogen peroxide over Fe/activated carbon catalyst*, J. Mol. Catal. A: Chem. 272 **(2007)** 169-173.
50. M. Schwidder, S. Heikens, A. D Toni, S. Geisler, M. Berndt, A. Bruckner, W. Grunert, *The role of NO2 in the selective catalytic reduction of nitrogen oxides over Fe-ZSM-5 catalysts: Active sites for the conversion of NO and of NO/NO2 mixtures*, J. Catal. 259 **(2008)** 96-103.
51. M. Santhosh Kumar, M. Schwidder, W. Grünert , A. Brückner, *On the nature of different iron sites and their catalytic role in Fe-ZSM-5 DeNOx catalysts: new insights by a combined*

EPR and UV/VIS spectroscopic approach, J. Catal. 227 **(2004)** 384397.
52. H. J. Wan, B. S. Wu, C. H. Zhang, H. W. Xiang ,Y. W. Li, B. F. Xu, F. Yi, *Study on Fe-Al2O3 interaction over precipitated iron catalyst for Fischer-Tropsch synthesis,* Catal. Commun.8 **(2007)** 1538-1545.
53. H. Y. Lin, Y. W. Chen, C. Li, *The mechanism of reduction of iron oxide by hydrogen*, Thermochim. Acta 400 **(2003)** 61-67.
54. Y. Jin and A. K. Datye, *Phase Transformations in Iron Fischer-Tropsch Catalysts during Temperature-Programmed Reduction*, J. Catal. 196 **(2000)** 8-17.
55. G. Neria, A. M Visco, S. Galvagno, A. Donato, M. Panzalorto, *Au/iron oxide catalysts: temperature programmed reduction and X-ray diffraction characterization*, Thermochimi. Acta 329 **(1999)** 39-46.
56. L.A. Boot, A.J.vanDillen, J.W.Geus, F.R.van Buren, *Iron-Based Dehydrogenation Catalysts Supported on Zirconia. I. Preparation and Characterization*, J. Catal. 163 **(1996)**186-194.
57. F.G.E. Nogueira, J.H. Lopes, A.C. Silva, R.M. Lago, J.D. Fabris, L.C.A. Oliveira, *Catalysts based on clay and iron oxide for oxidation of toluene*, Appl. Clay Sci. 51 **(2011)** 385-389.
58. T. V. Barakat, V. Idakiev, R. Cousin, G. S. Shao, Z. Y. Yuan, T. Tabakova, S. Siffert, *Total oxidation of toluene over noble metal based Ce, Fe and Ni dopedtitanium oxides*, **(2013)** http://dx.doi.org/10.1016/j.apcatb.2013.05.064 .
59. H. C. Wang, H. S. Liang, M. B. Chang, *Chlorobenzene oxidation using ozone over iron oxide and manganese oxide catalysts,* J. Hazard. Mater.186 **(2011)** 1781-1787.
60. S. S. Masiero, N. R. Marcilio, O. W. Perez-Lopez, *Aromatization of Methane Over Mo-Fe/ZSM-5 Catalysts*, Catt. Lett 131 **(2009)** 194-202.
61. B. Shen, T. Liu, N. Zhao, X. Yang, L. Deng, *Iron-doped Mn-Ce/TiO2 catalyst for low temperature selective catalytic reduction of NO with NH3*, Journal of Environmental Sciences 22(9) **(2010)** 1447-1454.
62. Y. Wang, Q. Zhang, T. Shishido,and K. Takehira, *Characterizations of Iron- Containing MCM-41 and Its Catalytic Properties in Epoxidation of Styrene with Hydrogen Peroxide*, J. Catal. 209 **(2002)** 186-196.
63. R. Q. Long and R. T. Yang, *Temperature-Programmed Desorption/Surface Reaction (TPD/TPSR) Study of Fe-Exchanged ZSM-5 for Selective Catalytic Reduction of Nitric Oxide by Ammonia*, J. Catal. 198 **(2001)** 20-28.
64. A. V. Vijayasankar, C. U Aniz and N. Nagaraju, *Surface Properties and the Catalytic Activity of Amorphous Iron Aluminophosphates: Effect of Fe Loading*, J. Korean Chem. Soc. 54 **(2010).** DOI 10.5012/jkcs.2010.54.01.131.
65. N. E. Bogdanchikova, M. N. Dulin, A. V. Toktarev, G. B. Shevnina, V. N. Kolomiichuk, V. I Zaikovskii, V. P. Petranovskii, *Stabilization of silver clusters in zeolite matrices* ,Stud. Surf. Sci. 84 **(1994)** 1067-1074.
66. P. Boron, L. Chmielarz, S. Dzwigaj, *Influence of Cu on the catalytic activity of FeBEA zeolites in SCR of NO with NH3*, Appl. Catal. B: Environ.168-169 **(2015)** 377-384.
67. Z. Xian-wei, S. Ya-xin, C. Jiang-hao, L. Rui , W. Ni-ni, D. Wen-yi , Z. Hao, *Effect of Ag on deNOx performance of SCR-C3H6 over Fe/Al-PILC catalysts*, J Fuel Chem Technol, 47 **(2019)** 1368-1378.
68. X. Zhang, Y. Yang, L. Song, Y. Wang, C. He, Z. Wang, L. Cui, *High and stable catalytic activity of Ag/Fe2O3 catalysts derived from MOFs for CO oxidation*, Mol. Catal. 447 **(2018)** 80-89.

Chapter IV

Catalytic activity of materials in toluene oxidation

IV. 1 Introduction

In the literature review on the oxidation reaction of VOCs (chapter I) we reported the oxidation mechanism of VOCs on different types of catalyst (supported or unsupported). This mechanism, proposed by Mars and Van Krevelen, is of the redox type [1-6] and is based on a cycle of oxygen species that move between the support and the ambient gas. It specifies that the reactant is oxidised by an oxygen in the support. The oxidised support then reduces and forms a vacant site in its matrix, which is occupied again by another oxygen atom thanks to a reaction between the support and the gaseous oxygen [7].

It is important to note that work reported in the literature has shown that in addition to the redox properties of the catalyst and the high mobility of oxygen on its surface, its acidic properties are necessary for its catalytic activity in the oxidation of VOCs [16, 8-13]. The presence of acidic sites enables the VOC to be sorbed to the surface of the catalyst and the carbon-carbon bonds to be broken to produce CO_2 [14].

In this chapter we present the results of the analysis of the catalytic activity of the solids M-ZrO2 and Ag/M-ZrO2 in the oxidation reaction of toluene in the presence of H_2O between 200 and 550°C. The effects of the nature and quantity of M (M= Ce or Fe) and the silver content were studied. An explanation of the catalytic properties as a function of the physicochemical characteristics is reported.

IV. 2 Catalytic activity of solids y Ce-ZrO2 and Ag/yCe- ZrO2 in toluene oxidation.

IV.2.1 Catalytic activity of Ce-ZrO2 supports: effect of Ce content

The results of the changes in toluene (C_6H_5-CH_3) conversion and CO_2 concentration as a function of temperature on the catalyst supports (ZrO2 and yCe-ZrO2 with y =1.25; 2.5; 5; 10; 15; 20 and 30) are shown in Figures 50 and 51, respectively.

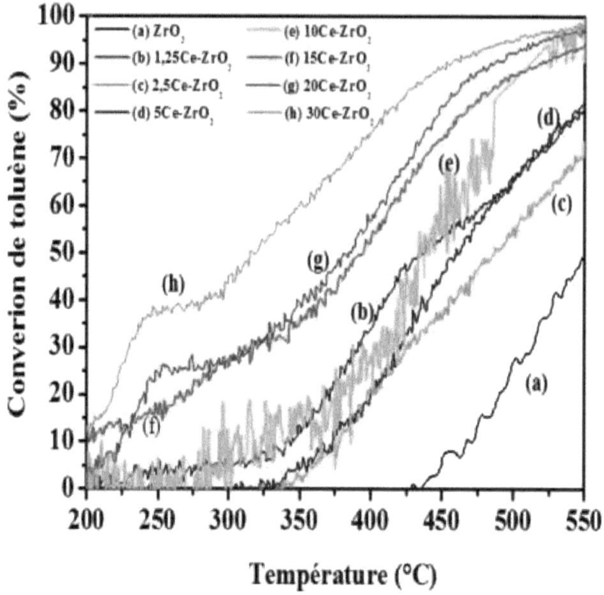

Figure 50: Conversion of toluene on yCe-ZrO2 catalyst supports.

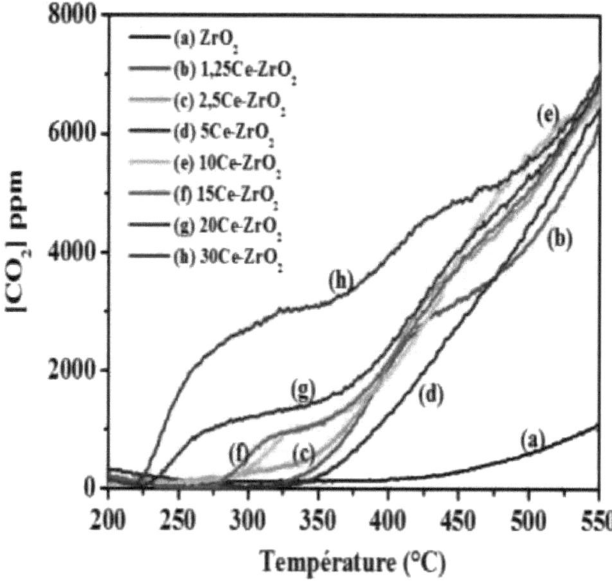

Figure 51. Concentration of CO_2 formed on the yCe-ZrO2 catalyst supports.

Figures 50 and 51 show that zirconia is not very active in the toluene (C_6H_5-CH_3) oxidation reaction. The reaction is characterised by a 50% conversion of C6H5-CH3 to CO_2 at 550°C. The addition of cerium increases the catalytic activity. The solids yCe- ZrO2 are more active than zirconia and their activity increases with the cerium content and the reaction temperature. The conversion of toluene to CO_2 is greater than 90% at 550°C in yCe-ZrO2 aerogels containing a cerium content of> 10% wt. This increase can be attributed, according to the results presented in chapter III, to the improvement in the mobility of surface oxygen and the acid and redox characteristics of the solids by the addition of cerium. Cerine is an "oxygen reservoir" and has a high oxygen transport capacity and an ability to move easily between its reduced and oxidised forms (Ce^{3+}- Ce^{4+}), which gives it a tendency towards strong catalytic activities. Numerous studies have shown that the presence of cerium facilitates the creation of oxygen vacancies on the surface, promotes the creation of new active sites and improves the activity of materials in the C6H5-CH3 oxidation reaction [15-20].

The highest toluene conversion rates are obtained on solids containing large amounts of Ce (% Ce > 10%) but when the amount of cerium exceeds the value of 10%, the CO_2 selectivity decreases and a low intensity peak, relating to the fragment m/e = 43 and attributed to a compound of the C3-C6 paraffin family [21], is observed. This difference in the selectivity of the solids may be linked to the change in the nature of the cerium species, acid sites and redox sites as the amount of Ce increases.

IV.2.2 Catalytic activity of Ag/yCe-ZrO2 solids

The results of the toluene oxidation catalytic test at 200 to 550°C (C6H5-CH3 conversion and CO_2 concentration) on Ag/yCe-ZrO2 catalysts (y =1.25; 2.5; 5; 10; 15; 20 and 30) are shown in Figures 52 and 53, respectively.

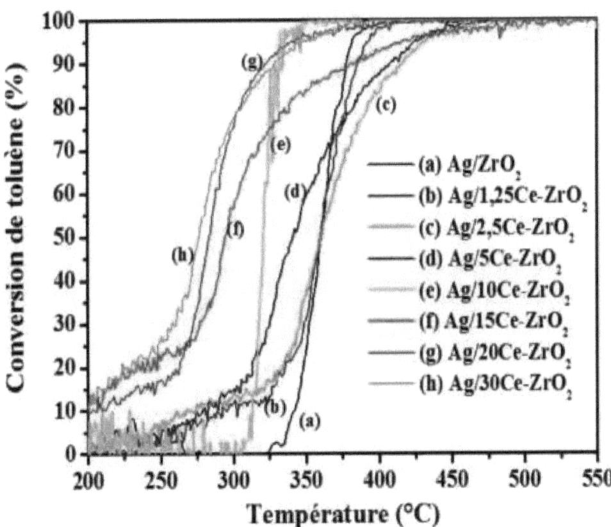

Figure 52. Conversion of toluene over Ag/yCe-ZrO2 catalysts.

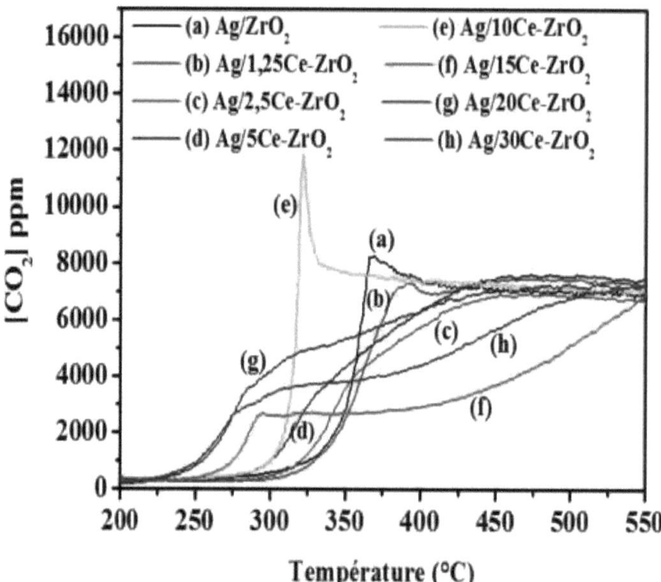

Figure 53. Concentration of CO_2 formed on Ag/yCe-ZrO2 catalysts.

Figure 52 shows good catalytic activity of all the Ag/yCe- ZrO2 samples in toluene oxidation. The conversion of C6H5-CH3 increases with increasing Ce content and is significantly greater in the case of the Ag/yCe-ZrO2 catalysts compared with the yCe-ZrO2 supports (Fig 50). This result shows that silver is very active in the oxidation of toluene and is in agreement with several studies which have also reported an increase in catalytic activity after the addition of silver to different supports such as CeO_2, ZSM-5 and a-MnO2 [22-25].

The good catalytic activity of the Ag-containing catalysts can be explained by the presence of new acid and redox sites created by the silver and probably also by the good mobility of the surface oxygen due to the interactions between the silver and cerium species. Z. Qu et al [26] studied the effect of the addition of Ag on Mn/SBA-15 solids and showed that the interactions between Ag and Mn improve the redox properties of Ag-Mn/SBA-15 solids and the mobility of oxygen and therefore influence the catalytic activity of these materials in toluene oxidation.

The results also showed that the CO_2 selectivity of the Ag/yCe-ZrO2 catalysts depends on the cerium content. The Ag/10Ce-ZrO2 solid gives total selectivity in CO_2 between 200 and 550°C: it is the best performing catalyst with 100% conversion of C6H5-CH3 with 100% selectivity in CO_2 between 350 and 550°C. For the other catalysts (Ag/1.25Ce-ZrO2; Ag/2.5Ce-ZrO2; Ag/5Ce-ZrO2; Ag/15Ce-ZrO2; Ag/20Ce-ZrO2; Ag/30Ce-ZrO2), small quantities of benzene and a compound from the C3-C6 paraffin family were detected at > 300°C.

Taking into account the physico-chemical properties of the Ag/10Ce-ZrO2 catalyst, we can deduce that its good catalytic activity is probably due to the presence of Ag_n^s clusters+ which absorb around 390 nm and/or Ag_n metallic silver aggregates which are not formed on the surface of the other Ag/yCe-ZrO_2 solids (y Φ 10).

IV.2.3 Effect of Ag content on the catalytic activity of xAg/Ce-ZrO2 solids

In order to gain a better understanding of the influence of Ag on the catalytic activity of x Ag/Ce-ZrO2 solids in the oxidation of C6H5-CH3, we tested in this reaction a series of catalysts containing increasing amounts of Ag (0.5, 1, 2 and 3% wt.). The results obtained are shown in Figures 54 and 55.

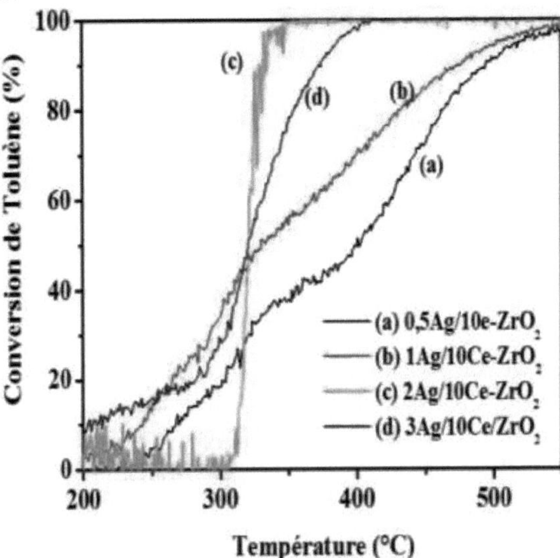

Figure 54. Conversion of toluene over xAg/10Ce-ZrO2 catalysts.

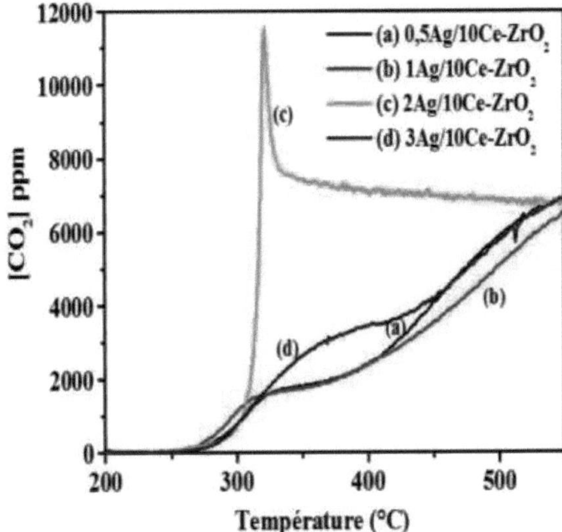

Figure 55. co2 concentration on xAg/Ce-ZrO2 catalysts.

Figure 54 shows that the catalytic activity in toluene oxidation of the xAg/Ce-ZrO2 catalysts increases with increasing Ag content up to an amount of 2% wt. The conversion of C6H5-CH3 decreases between 300 and 400°C for a higher amount of Ag. Note that 100% conversion of toluene to co2 is achieved at T > 300°C and T > 400°C on the 2Ag/Ce-ZrO2 and 3Ag/Ce-ZrO2 catalysts, respectively. Thus, 2% wt. Ag is the optimum amount of silver for best activity in toluene oxidation for T> 300 °C.

Taking into account the results of the nitrogen physisorption analysis (Chapter III), we can link the better catalytic performance of the 2Ag/10Ce-ZrO2 sample to the presence of Ag_n^s clusters+ and Ag_n metallic silver aggregates and to its large specific surface area (73 $m^2.g^{-1}$), compared with the solid 3Ag/10Ce-ZrO2 (55.8 $m^2.g^{-1}$)'which allows good dispersion of these species. On the other hand, the slight decrease in catalytic activity observed on the solid containing 3% wt Ag can be attributed to the formation of metallic silver on the surface of this catalyst. Indeed, many studies [1-2,5] have attributed the decrease in the catalytic activity of Ag-containing catalysts to a significant agglomeration of silver species on the surface of the supports. Other studies [4,27-32] have revealed the important role of good dispersion of the active species on the catalyst surface in the VOC oxidation reaction. M. Skaf et al [33] concluded that the large particles of the catalyst lose their contact with the support and the probability of formation of the surface oxygenated species which are necessary for the oxidation of the VOCs.

We can conclude that the quantities of cerium and silver influence the physicochemical properties of the Ag/Ce-ZrO2 catalysts and thus affect their catalytic performance in the oxidation of toluene. Better catalytic activity was obtained on the catalyst containing 10% wt Ce and 2% wt Ag. 100% conversion of C6H5-CH3 to CO2 was obtained on this catalyst between 350 and 550°C.

IV. 3 Catalytic activity of Fe-ZrO2 and Ag/Fe-ZrO2 solids in toluene oxidation.

IV.3.1 Catalytic activity of Fe-ZrO2 supports: Effect of Fe content on the catalytic activity of Fe-ZrO2 supports

The changes in C6H5-CH3 conversion and CO2 concentration as a function of reaction temperature on the yFe-ZrO2 catalyst supports (containing increasing amounts of Fe; y =1.25, 2.5, 5 and 10) are shown in Figures 56 and 57, respectively.

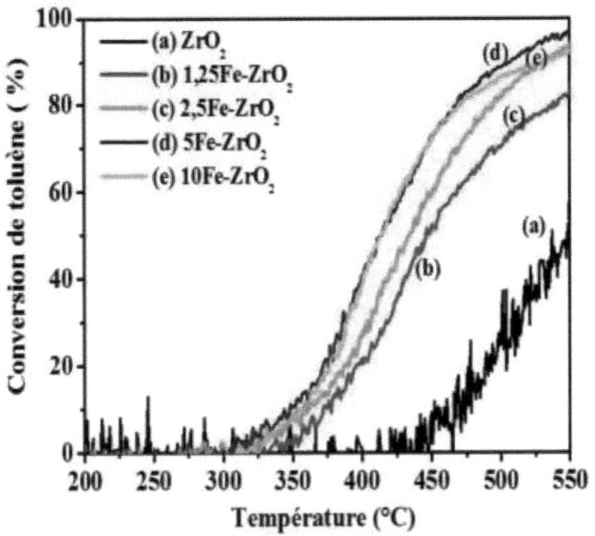

Figure 56. Conversion of toluene on yFe-ZrO2 catalyst supports.

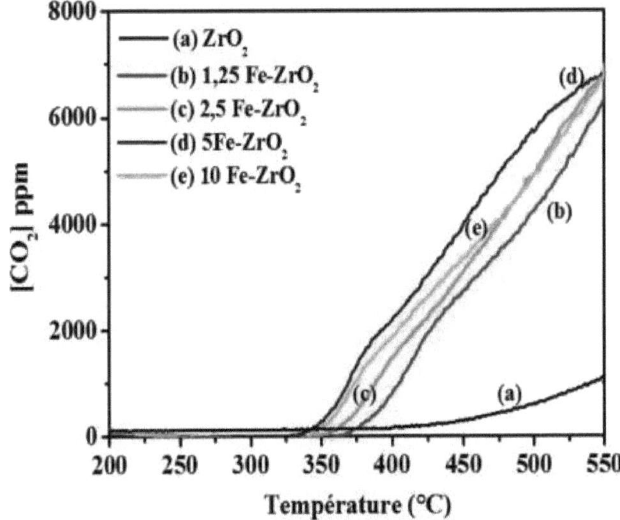

Figure 57. Concentration of CO2 formed on the yFe-ZrO2 catalyst supports.

Analysis of this figure shows that yFe-ZrO2 aerogels are more active in toluene oxidation than pure zirconia. This improvement in catalytic activity is attributed to the creation of new acid and redox sites by the addition of iron species to the zirconia surface. It is also probably due to an increase in the mobility of surface oxygen, which favours the toluene oxidation reaction [13,14,18].

It should be noted that the activity of the yFe-ZrO2 aerogels is affected by the amount of iron added. In fact, the best catalytic activity (conversion of C6H5-CH3 and selectivity to CO2) is observed in the case of the sample containing 5% wt.Fe, which gives a maximum toluene conversion of around 95% at 550°C. For a higher Fe content (10% wt), a slight decrease in catalytic activity was noted. The effect of the amount of iron on the catalytic activity of several solids has been studied in the literature. P. Djinovic et al [12] showed that the amount of Fe influences the mobility and reactivity of the oxygen species and therefore modifies the catalytic activity of the CuFe bimetallic system in toluene oxidation. On the other hand, X. Liang et al [14] found that the activity of Fe-Ti-PILCs catalysts in the same reaction increased with increasing iron content up to 3.75%.

IV. 3.2 Catalytic activity of Ag/yFe-ZrO2 solids

The changes in C6H5-CH3 conversion and CO_2 concentration as a function of temperature over the Ag/yFe-ZrO2 catalysts are shown in Figures 58 and 59.

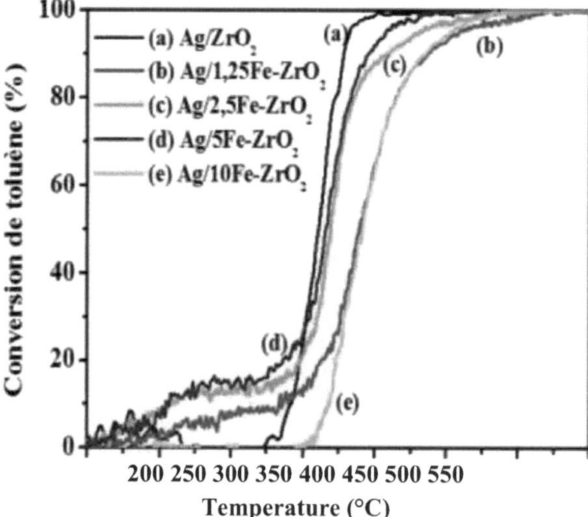

Figure 58. Conversion of toluene over Ag/yFe-ZrO2 catalysts.

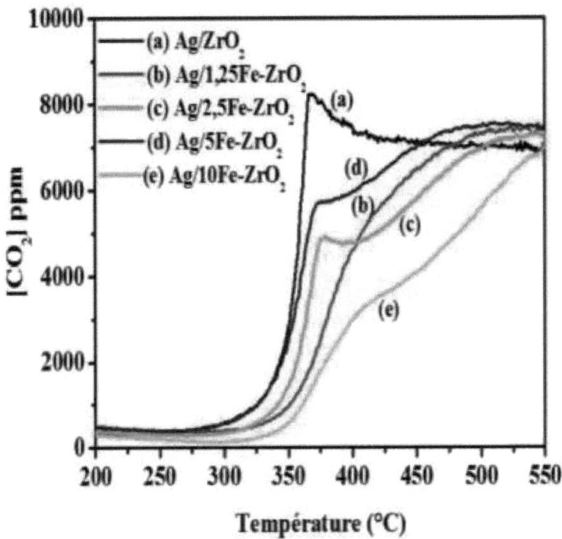

Figure 59. Concentration of CO_2 formed on Ag/yFe-ZrO_2 catalysts.

A comparison of the results obtained in Figures 58 and 59 shows a significant increase in toluene conversion after the addition of silver to the yFe-ZrO_2 supports. This result reveals that the silver species are more active in the oxidation of toluene than the iron species. Furthermore, the catalytic activity increases in the following order: Ag/1.25Fe-ZrO2 ~ Ag/10 Fe-ZrO2< Ag/2.5Fe-ZrO2< Ag/5Fe-ZrO2~ Ag/ZrO2. The Ag/5Fe-ZrO2 and Ag/ZrO2 samples are therefore the most active solids in this reaction; they show almost similar catalytic powers at high temperature with almost 100% conversion of toluene to CO_2 for T > 400°C (Figures 58 and 59). In contrast, the Ag/5Fe-ZrO2 solid is the most active at low temperatures (T < 225°C). This may be related to its higher redox power and probably also to an increase in the mobility of the surface oxygen due to the existence of interactions between the iron and silver species, which improve the redox characteristics of the catalysts and increase the oxidation activity of toluene [34-36].

IV.2.3 Influence of Ag content on the catalytic activity of xAg/Fe-ZrO2 solids

The effect of Ag content (from 0.5 to 3%) on the catalytic activity of xAg/Fe- ZrO2 solids is studied and the results obtained are illustrated in Figures 60 and 61.

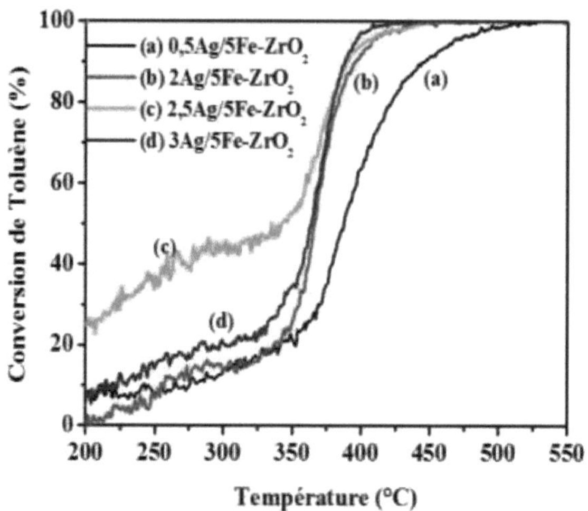

Figure 60: Conversion of toluene over xAg/5Fe-ZrO2 catalysts.

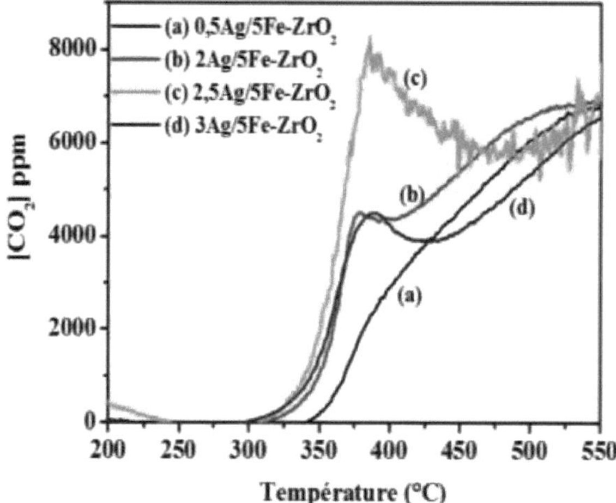

Figure 61. Concentration of CO_2 formed as a function of temperature on xAg/5Fe-ZrO2 catalysts.

Analysis of the results illustrated in Figures 60 and 61 shows an improvement in toluene conversion as the Ag content increases up to a quantity of 2.5% wt. For a higher amount of Ag the catalytic activity decreases, particularly at low temperatures (T < 350°C). Thus we can deduce that 2.5% wt Ag is the optimum amount of silver for the synthesis of the 2.5Ag/5Fe-ZrO2 catalyst, the most active in the oxidation of toluene to CO_2 in this series. From the data in Table 24 (Chapter III), we can suggest that the good specific surface area of this catalyst allows good dispersion of the active species, which leads to excellent activity in toluene oxidation. The decrease in catalytic activity for a 3% wt Ag content can be explained by the formation of metallic silver on the surface of the 3 Ag/5Fe-ZrO2 solid, which probably reduces the number of active sites and catalytic activity.

We can conclude that the quantities of iron and silver influence the physicochemical properties of the Ag/Fe-ZrO2 catalysts and affect their catalytic activities in the oxidation of toluene. The catalyst containing 5% wt Fe and 2.5% wt Ag is the most active, with 100% conversion of toluene essentially to CO_2 being obtained on this catalyst between 400 and 550°C.

IV 3 Comparative study of the catalytic power of our catalysts in the oxidation of toluene with those tested in the literature

To date, no work has been done in the literature on the oxidation reaction of toluene over silver-based catalysts supported on aerogel supports (ZrO_2, Ce-ZrO2 and Fe-ZrO2). In this work we focused on the synthesis and characterisation of new catalysts (Ag/ZrO2, Ag/Ce-ZrO2 and Ag/Fe-ZrO2) for the toluene oxidation reaction. The results obtained on some of our catalysts (the most active) and those reported in the literature for other catalysts are shown in Tables 26 and 27. This study shows that our results are interesting due to the fact that the reaction is carried out in our study under harsher conditions than those reported in the literature, which are as follows: the presence of H_2O (3.4%), a high gas space velocity (60,000 h^{-1}) and a low catalyst mass (m = 20 mg).

Table 26. Comparison of the catalytic activity in toluene oxidation of our catalysts and those reported in the bibliography.

Catalyst	T10 (°C)	T50 (°C)	T90 (°C)	Selectivity	Catalytic test conditions	Preparation method	Ref
10Ce-ZrO2	327	436	511	CO_2	GHSV = 60,000 h^{-1}, m = 20 mg with 3.4 Vol % H2O	Sol gel	Our study
CeO2	~160	~230	-	Benzyl alcohol Benzoic acid CO_2	GHSV = 22500 h^{-1}, m = 200 mg Without H2O	Precipitation	[37]
CeO2/SBA-15	~277	~352	-	CO_2	GHSV = - m= 30 mg without H_2O	-	[38]
2Ag/10Ce-ZrO2	312	320	330	CO_2	GHSV = 60,000 h^{-1}, m = 20 mg with 3.4 Vol % H2O	substrate: Sol gel catalyst: impregnation	Our study
Ag/YAl2O3	~330	~360	~380	CO_2	GHSV = 15,000 h^{-1} m = 200 mg without H_2O	Support : commercial Catalyst : Impregnation	[39]
1Ag/Y-Al2O_3	~400	-	-	CO_2	GHSV = 15,000 h^{-1}, m = 200 mg without H_2O	Support : commercial Catalyst : Impregnation	[40]
11Ag/1Cu- Y-Al2O3	~230	~250	~260	CO_2	GHSV = 15,000 h^{-1} m = 200 mg without H_2O	Support : commercial Catalyst : Impregnation	[39]
0.81Ag/3DOM 26.9CeO2-Al2O_3	263	308	338	---	GHSV = 20,000 ml/g h, m = 50 mg without H_2O		[41]

Table 27. Comparison of the catalytic activity in toluene oxidation of our catalysts and those reported in the bibliography.

Catalyst	T10 (°C)	T50 (°C)	T90 (°C)	Selectivity	Catalytic test conditions	Preparation method	Ref
5Fe-ZrO2	349	414	506	CO_2 C3-C6	GHSV = 60,000 h^{-1}, m =	Sol gel	Our

				paraffins	20 mg with 3.4 Vol % H2O		study [42]
Fe-TiO2	~450	-	-	CO2	GHSV = 60,000 h^{-1}, m = 100 mg No H_2O	Impregnation	[13]
Fe2O3	~325	~450	-	CO2	m = 30 and 50 mg Without H_2O	commercial	[42][4 3]
Fe-imp	~325	~350	~380	CO2	m = 30 and 50 mg Without H_2O	Impregnation	[41]
CuFe-SBA-15	~252	~327	-	CO2	GHSV = 1.2 h^{-1}, m = 30 mg No H_2O		[43][4 4]
2.5Ag/5Fe-ZrO2	200	~350	~394	CO2 C3-C6 paraffins	GHSV = 60,000 h^{-1}, m = 20 mg with 3.4 Vol % H2O	substrate: Impregnation catalyst gel sol	Our study
Ag-V/Ti	~342	~386	-	Benzyl alcohol Benzoic acid CO2 CO	GHSV = 31,000 h^{-1}, m = 100 mg No H_2O	----	[44] [45]
10Ag-U	~265	~280	~295	CO2	GHSV =30.000 ml/g/h, m = 100 mg without H_2O	Liquid phase reduction method	[45] [46]
10Ag-U	267	286	300	CO2	GHSV = 30,000 ml/g/h m = 100 mg 10% H2O	Liquid phase reduction method	[44] [45]
Ag/SBA-15	~255	~270	~280	CO2	GHSV = 15,000 h^{-1}, m = 200 mg without H_2O	Impregnation	[46] [47]

IV. 4 Conclusions

The results obtained in this chapter show that :

- Pure zirconia is characterised by very low catalytic activity in the oxidation reaction of toluene (C6H5-CH3). The addition of Ag, Fe or Ce improves catalytic activity by improving the acid and redox properties of the solids and increasing the mobility of surface oxygen.
- Silver species are more active in the toluene oxidation reaction than cerium or iron species.
- The activity of Ce/ZrO2 or Fe/ZrO2 aerogel catalyst supports is affected by the quantities of Fe (1.25; 2.5; 5 and 10%) and Ce (1.25; 2.5; 5; 10; 15; 20 and 30%). The quantities 10% wt. Ce and 5% wt. Fe lead to the most active catalyst supports for toluene oxidation.

• Ce/ZrO2 allows 100% conversion of toluene almost completely into CO2 at 550°C.

• FE/ZRO2 converts 95% of toluene almost completely to CO2 at 550°C.

• 2AG/ZRO2 enables 100% conversion of toluene to CO2 in the 350-550°C temperature range.

- Varying the amounts of Ce (1.25; 2.5; 5; 10; 15; 20 and 30%) and Fe (1.25; 2.5; 5 and 10%) influences the catalytic activity of the 2Ag/yCe- ZrO2 and 2Ag/yFe-ZrO2 catalysts.

- Ag/10 Ce-ZrO2 and Ag/5Fe-ZrO2 are the most active:

• Ag/10 CE-ZRO2 enables 100% conversion of C6H5-CH3 to CO2 between 350 and 550°C.

• AG/5FE-ZRO2 allows 100% conversion of C6H5-CH3 at T > 400°C.

- Varying the amount of silver (0.5, 1, 2, 2.5 and 3%) modifies the catalytic power of the Ag/10Ce-ZrO2 and Ag/5Fe-ZrO2 catalysts in the oxidation of toluene. The use of the 2 and 2.5% Ag contents enables the preparation of the 2Ag/10 Ce-ZrO2 and 2.5Ag/5Fe-ZrO2 catalysts that are most active in this reaction.

- 2.5AG/5FE-ZRO2 allows 100% conversion of toluene essentially to CO2 between 350 and 550°C.

IV.5 bibliographical references

1. S. Todorova, A. Naydenov, H. Kolev, J. P. Holgado, G. Ivanov, G. Kadinov, A. Caballero, *Mechanism of complete n-hexane oxidation on silica supported cobalt and manganese catalysts.* Appi Catal A Gen 413- 414 **(2012)** 43-51.

2. A. k. Neyestanaki, N. Kumar, L. E. Linfors, *Catalytic combustion of propane and natural gas over Cu and Pd modified ZSM zeolite catalysts*. Appl Catal B 7 **(1995)** 95111.

3. H. Huang, Y. Xu, Q. Feng, D.Y.C. Leung, *Low temperature catalytic oxidation of volatile organic compounds: a review*. Catal Sci Technol 5 **(2015)** 2649-2669.

4. J.C. Menezo, J. Riviere, J. Barbier, *Effect of the doping of a metal-oxide by platinum on its oxidizing properties*. React Kinet Catal Lett 49 **(1993)** 293-298.

5. Y. Qin, Z. Qu, C. Dong, N. Huang, *Effect of pretreatment conditions on catalytic activity of Ag/SBA-15 catalyst for toluene oxidation*. Chinese J Catal. 38 **(2017)** 16031612.

6. J. Li, Z. Qu, Y. Qin, H. Wang, *Effect of MnO2 morphology on the catalytic oxidation of toluene ove rAg/MnO2 catalysts*. Appl Surf Sci 385 **(2016)** 234-240.

7. P. Mars, D.W. Van Krevelen, *Oxidations carried out by means of vanadium oxide catalysts*, Chem. Eng. Sci3 **(1954)** 41-59.

8. M. Popova, A. Szegedi, Z. Cherkezova-Zheleva, A. Dimitrova, I. Mitov, *Toluene oxidation on chromium- and copper-modified SiO2 and SBA-15*, Appl. Catal. A: Gen. 381 **(2010)** 26-35.

9. P. Papaefthimiou, T. Ioannides, X.E. Verykios, *VOC removal: investigation of ethylacetate oxidation over supported Pt catalysts*, Catal. Today, 54 **(1999)** 81-92.

10. C. Lamonier, J. F Lamonier, B. Aellach, A. Ezzamarty, J. Leglise, *Specific tuning of acid/base sites in apatite materials to enhance their methanol thiolation catalytic performances,* Catal. Today 164 **(2011)** 124-30.

11. M. F. Finol. J. Q. Torres, J. M. Giraudon., A. Gervasini, J. F Lamonier, *Formaldehyde oxidation over hydroxyapatites: influence of the acid base properties,* Europa Cat X, Glas gow (Scotland) **(2011)**.

12. P. Djinovic, A. Ristic , T. Zumbar, V. D. B. C. Dasireddy, M. Rangus, G. Drazic, M. Popova, B. Likozar, N. Z. Logar and N. N. Tusar. *Synergistic Effect of CuO Nanocrystals and Cu-oxo-Fe Clusters on Silica Support in Promotion of Total Catalytic Oxidation of Toluene as a Model Volatile Organic Air Pollutant.* Appl. Catal. B. Environ. **(2020)** https://doi.org/10.1016/j.apcatb.2020.118749

13. X. Liang , F. Qi , P. Liu, G. Wei , X. Su, L. Ma, H. He, X. Lin, Y. Xi, J. Zhu, R. Zhu, Appl .Clay. Sci. *Performance of Ti-pillared montmorillonite supported Fe catalysts for toluene oxidation: The effect of Fe on catalytic activity.* **(2016)** http://dx.doi.org/10.1016/j.clay.2016.05.022

14. H. Windawi, M. Wyat , *Catalytic destruction of halogenated VOC over platinum group metal catalysts*, Platinum Metals Rev 37 **(1993)** 186-248.

15. E. Aneggi, J. Lorca, Carla de Leitenburg , G. Dolcetti, A. Trovarelli, *Soot combustion over silver-supported catalysts.* Appi. Catal. B. Environ. 91 **(2009)** 489-498.

16. Z. Guerra-Que, G. Torres-Torres, H. Pérez-Vidal, I. Cuauhtémoc-Lopez, A. Espinosa de los Monteros, J.N. Beltramini, D.M. Frias-M'arquez, *Silver nanoparticles supported on zirconia-ceria for the catalytic wet air oxidation of methyl tert-butyl ether*. RSC Adv 7 **(2017)** 3599-3610.

17. K. Yamazaki, Y. Sakakibara, F. Dong, H. Shinjoh, *The remote oxidation of soot separated by ash deposits via silver-ceria composite catalysts.* Appl Catal A Gen 476 **(2014)**

113-120.
18. K. Yamazaki, T. Kayama, F. Dong, H. A. Shinjoh, *mechanistic study on soot oxidation over CeO2-Ag catalyst with 'rice-ball' morphology,* J Catal 282 **(2011)** 289298.
19. Z. Qu , F. Yu, X. Zhang, Y. Wang, J. Gao, *Support effects on the structure and catalytic activity of mesoporous Ag/CeO2 catalysts for CO oxidation.* Chem Eng J 229 **(2013)** 522-532.
20. D. Valechha, S. Lokhande, M. Klementova, J. Subrt, S. Rayalu, N. Labhsetwar, *Study of nano-structured ceria for catalytic CO oxidation,* J Mater Chem 21 **(2011)** 37183725.
21. J. Brunet, E. Genty, Y. Landkocz, M. Al Zallouha, S. Billet, D. Courcot, S. Siffert , D. Thomas, G. D. Weireld, R. Cousin, *Identification of by-products issued from the catalytic oxidation of toluene by chemical and biological methods.* C. R. Chimie **(2015).** https://doi.org/10.1016/j.crci.2015.09.001
22. S. Imamura, H. I. Yamada, K. Utani, *Combustion activity of Ag/CeO2 composite catalyst,* Appl. Catal. A 192 **(2000)** 221-226.
23. S. Scirè, P. M. Riccobene, C. Crisafulli, *Ceria supported group IB metal catalysts for the combustion of volatile organic compounds and the preferential oxidation of CO,* Appl. Catal. B 101 **(2010)** 109-117.
24. A. Jodaei, A. Niaei, and D. Salari, *Performance of nanostructure Fe-Ag- ZSM-5 catalysts for the catalytic oxidation of volatile organic compounds: Process optimization using response surface methodology*, Korean J. Chem. Eng, 28(8) **(2011)** 1665- 1671.
25. Q. Ye, J. Zhao, F. I. Huo, J. Wang, S. Cheng, T. Kang, H. Dai, *NanosizedAg/a-MnO2 catalysts highly active for the low-temperature oxidation of carbon monoxide and benzene.* Catal. Today, 175 **(2011)** 603-609.
26. Z. Qu , Y. Bu, Y. Qin, Y. Wang, Q. Fu, *The improved reactivity of manganese catalysts by Ag in catalytic oxidation of toluene*, Appl Catal B: Environ132-133 **(2013)**353-362
27. T. Tabakova, F. Boccuzzi, M. Manzoli, J. W. Sobczak, V. Idakiev, D. Andreeva, *A comparative study of nanosized IB/ceria catalysts for low-temperature water-gas shift Reaction*, Appl. Catal. A: General, , Applied Catalysis A, 298 **(2006)** 127-143.
28. D. Chen, Z. Qu, S. Shen, X. Li, Y. Shi, Y. Wang, Q. Fu, J,Wu, *Comparative study of silver based catalysis supported on different supports for the oxidation of formaldehyde.* Catal. Today, 175 **(2011)** 338-345.
29. F. Boccuzzi, A. Chiorino, M. Manzoli, D. Andreeva, T. Tabakova, T. L Ilieva, V. Iadakiev. *Gold, silver and copper catalysts supported on TiO2 for pure hydrogen production.* Catal. Today 75 **(2002)** 169-175.
30. Q. Ye. J. Zhao, F.i Huo, J. Wang, S. Cheng, T. Kang, H. Dai. *NanosizedAg/a-MnO2 catalysts highly active for the low-temperature oxidation of carbon monoxide and benzene.* Catal. Today 175 **(2011)** 603-609.
31. E. M. Cordi, J. L. Falconer, *Oxidation of volatile organic compounds on an Ag/A1203 catalyst.* Appl. Catal **(1997)** 179-191.
32. N. Guilhaume, B. Bassou, G. Bergeret, D. Bianchi, F. Bosselet, A . Desmartin- Chomel, B ouguet, C. Mirodatos, *In situ investigation of Diesel soot combustion over an AgMnOx catalyst*, Appl. Catal. B 119-120 **(2012)** 287-296.
33. M. SKAF, Comparaison physico-chimique et des activités catalytiques dans les réactions d'oxydation, entre deux séries de catalyseurs Ag/CeO2 préparés par impégnation et dépôt précipitation, PhD thesis, **(2013)**, Université du Littoral côte d'opale and University of Balamand, France, Dubai.
34. L.J, Kundakovic, M. Flytzani-Stephanopoulos, *Deep oxidation of methane over zirconia*

supported Ag catalysts, Appl. Catal. A 183 **(1999)** 35-51.
35. A. Chongterdtoonskul, J. W. Schwank, S. Chavadej, *Effects of oxide supports on ethylene epoxidation activity over Ag-based catalysts.* J. Mol. Catal. A, 358 **(2012)** 5866.
36. R. B Grant, R. M Lambert, *A single crystal study of the silver-catalysed selective oxidation and total oxidation of ethylene,* J. Catal. 92 **(1985)** 364-375.
37. J. Du, Z. Qu, C. Dong, L. Song, Y. Qin, N. Huang, *Low-temperature abatement of toluene over Mn-Ce oxides catalysts synthesized by a modified hydrothermal approach*, Appl. Surf. Sci **(2017).** https://doi.org/10.1016/j.apsusc.2017.10.116
38. T. Tsoncheva, G. Issa, T. Blasco, M. Dimitrov, M. Popova, S. Hernandez, D. Kovacheva, G. Atanasova, M. Jose. L. Nieto, *Catalytic VOCs elimination over copper and cerium oxide modified mesoporous SBA-15 silica.* Appl. Catal. A: Gen 453 **(2013)** 1-12S.
39. C Kim, J. H Moon, *Effect of Controlling Nano-Sized Copper by Silver in CopperBased Catalyst on Catalytic Oxidation of Toluene*, Nanosci J Nanotechnol 1 **(2011)** 1660-1663
40. S.C. Kim, and J.Y Ryu, *Properties and performance of silver -based catalysts on the catalytic oxidation of toluene*, Environ. Technol. 32 **(2011)** 561-568
41. H. Yang, J. Deng, Y. Liu, S. Xie, Z. Wu, H. Dai, *Preparation and catalytic performance of Ag, Au, Pd or Pt nanoparticles supported on 3DOM CeO2-Al_2O_3 for toluene oxidation*, J. Mol. Catal. A: Chem. 414 **(2016)** 9-18
42. F. G.E. Nogueira, J. H. Lopes, A.C. Silva, R. M. Lago, J. D. Fabris, L. C.A. Oliveira, *Catalysts based on clay and iron oxide for oxidation of toluene*, Appl. Clay Sci. 51 **(2011)** 385-389
43. A. Szegedi, M. Popova, M. Lázár, S. Klébert, E. Drotár, *Impact of silica structure of copper and iron-containing SBA-15 and SBA-16 materials on toluene oxidation*, Microporous and Mesoporous Materials 177 **(2013)** 97-104.
44. H. Ge, G. Chen, Q. Yuan, H. Li, *Gas phase partial oxidation of toluene over modified V2O5/TiO2 catalysts in a microreactor*, Chem. Eng. J. 127 **(2007)** 39-46.
45. X.D Zhang, L. Song, F. Bi, D.F Zhang, Y.X Wang, L.F Cui, *Catalytic oxidation of toluene using a facile synthesized Ag nanoparticle supported on UiO-66 derivative*, J Colloid Interface Sci, 571 **(2020)** 38-47
46. Z. Qu, Y. Bu, Y. Qin, Y. Wang, Q. Fu, *The improved reactivity of manganese catalysts by Ag in catalytic oxidation of toluene*, Appl. Catal. B: Environ. 132-133 **(2013)** 353-362.

General conclusion

In this work, silver-based catalysts supported on ZrO2 and M-ZrO2 supports, with M = Ce and Fe, were prepared for the toluene oxidation reaction. The effects of the content of M and the amount of silver on the physicochemical and catalytic properties of the materials obtained are studied.

The results of the physico-chemical analyses showed that:

> The zirconia, synthesised by the Sol-Gel method, is characterised by a good texture (S_{BET} = 107 $m^2.g^{-1}$ and Vp = 0.33 $cm^3.-g^{-1}$) and a structure containing the stable monoclinic and metastable tetragonal phase of ZrO2.

> The nature of M and its content (Ce (1.25; 2.5; 5; 10; 15; 20 and 30%) and Fe (1.25; 2.5; 5 and 10%)) influence the physicochemical properties of the Ce-ZrO2 or Fe-ZrO2 solids and the dispersion of the Ce or Fe species. The addition of Ce or Fe stabilises the tetragonal phase of ZrO2 (metastable at low temperatures) by improving its acid and redox properties. In addition, Ce-ZrO2 or Fe-ZrO2 solids retain a high specific surface area and porosity after the addition of an increasing quantity of M, reflecting the good thermal stability of these solids.

> The addition of increasing quantities of silver (x= 0.5; 1, 2; 2.5 and 3 %) leads, on the one hand, to a reduction in the specific surface area and pore volume of the Ce-ZrO2 or Fe-ZrO2 supports and, on the other hand, to the creation of new acid and redox sites on the surface of the Ag/Ce-ZrO2 and Ag/Fe-ZrO2 catalysts. The variation in the amount of silver also influences the nature of the silver species and their dispersion on the surface of the yCe-ZrO2 or yFe-ZrO2 aerogels.

The results of the total combustion reaction of toluene revealed that:

> Zirconia is not very active in toluene oxidation (50% conversion of C6H5-CH3 to CO2 at 550°C). On the other hand, the addition of Ce or Fe leads to Ce-ZrO2 or Fe-ZrO2 aerogel supports that are much more active than ZrO2. The activity of these supports is influenced by the nature of M (M= Ce or Fe) and its content; Ce (1.25; 2.5; 5; 10; 15; 20 and 30%) and Fe (1.25; 2.5; 5 and 10%):

- Ce-ZrO2 achieves 100% conversion of toluene almost completely to CO2 at 550°C.
- FE-ZRO2 achieves 95% conversion of toluene almost entirely to CO2 at 550°C.
- The quantities 10% wt. Ce and 5% wt. Fe lead to the most active and selective catalyst supports for toluene oxidation.

> The addition of silver improves catalytic activity of the catalyst supports in the oxidation of C6H5-CH3 and leads to Ag/ZrO2 or Ag/M- ZrO2 catalysts that are very active in this reaction. The results also showed that Ag species are more active than Ce and Fe species in the oxidation of toluene.

> The toluene conversion rate and the nature of the products formed depend on the nature of the metal M, its content and the quantity of silver :

- AG/ZRO2 enables 100% conversion of toluene to CO2 in the 350-550°C temperature range.
- The AG/10CE-ZRO2 and AG/5FE-ZRO2 catalysts are the most active and selective in the toluene oxidation reaction. Ag/10 Ce- ZrO2 enables 100% conversion of C6H5-CH3 to CO2 between 350 and 550°C and Ag/5Fe-ZrO2 enables 100% conversion of C6H5-CH3 at T > 400°C.
- 2 and 2.5% Ag are the optimum contents for preparing the 2Ag/10 Ce-ZrO2 and 2.5Ag/5Fe-ZrO2 catalysts that are most active in the toluene oxidation reaction.
-

Printed by Books on Demand GmbH, Norderstedt / Germany